$6.95

Your Environment And What You Can Do About It

By Richard Saltonstall, Jr.

It is *your* environment that everyone is talking about. It is your air that is dangerous to breathe, your water that is filled with effluents, your ears that are assailed with noise. But what can you do about it? Aren't the problems so vast and the solutions on such a massive and expensive scale as to preclude a role for the individual citizen?

In this information-packed book, Richard Saltonstall, Jr. provides a prescription for effective citizen action. Here, in readily usable form, is all the basic information on environmental quality that every citizen needs at his fingertips. And here, for the first time, is a thorough report on successful efforts made by citizens and citizen groups throughout the nation to solve the problems of their environment.

In short, this book tells you what you can do about air and water pollution, noise, chemical pollution of the land, urban sprawl, clogged highways, rural decay, waste disposal, and litter. It tells you what organizations you can apply to for more information, which government agencies to approach, and what your legal rights are.

It *is* your environment, after all, and you should play a role in its future—and yours.

YOUR ENVIRONMENT

And What You Can Do About It ❧❧❧❧❧❧❧❧❧❧

YOUR

And What You Can

Richard Saltonstall, Jr.

ENVIRONMENT
Do *About It* ✻✻✻✻✻✻✻✻✻✻✻✻✻✻

WALKER AND COMPANY

NEW YORK

FIRST PUBLISHED IN THE UNITED STATES OF AMERICA IN 1970 BY
THE WALKER PUBLISHING COMPANY, INC.
PUBLISHED SIMULTANEOUSLY IN CANADA BY THE RYERSON
PRESS, TORONTO.
LIBRARY OF CONGRESS CATALOG CARD NUMBER: 75–126112
PRINTED IN THE UNITED STATES OF AMERICA.
ISBN: 0–8027–0320–8

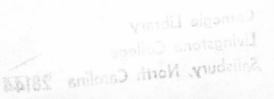

for my mother and father,
who taught me to appreciate the
unspoiled wonders of the land

RICHARD SALTONSTALL, JR. was an early advocate of the environment section in *Time* Magazine and his reporting has appeared regularly there since its inception. A graduate from Harvard in 1959, Mr. Saltonstall worked as a reporter for the Seattle *Times* and then joined Time, Inc. In addition to covering the conservation beat from the San Francisco and the Washington, D.C. offices of Time Inc. he has reported on a wide variety of subjects from sports to music to the American Indian.

CONTENTS

PREFACE

THIS BOOK COMES at a time when it is badly needed.

The year 1970 has seen a great change in the national purpose; in later years I believe we will look back and call it a watershed change. After 183 years of *development* as a national goal, in resources, land, family size, and standards of living, we appear to have permanently changed to a goal of *conservation*, directed at those resources of land, water, and air which contribute not only to our liberty and pursuit of happiness, but which are increasingly understood as being essential to life itself.

To say that preservation of our environment has only recently become a national priority is not without respect for those early conservationists who set aside our national parks and forests, or who worked to preserve vanishing species of fish and game. Nevertheless, despite the John Muirs, Gifford Pinchots, and others of their foresight, until now more conservation battles have been lost than won. Actions to save estuaries, redwoods, free-flowing streams, marshlands, and meadows have been fought all over the United States. Generally these battles were fought by a small group of local conservationists unable to obtain competent scientific or engineering testimony, unable to afford competent lawyers, and unable to hire high-priced public relations talent and media time. They have opposed the well-financed and well-entrenched power structure of local chambers of commerce and governmental officials who wanted to attract the payrolls of new industry, develop a new tax base, or construct public improvements at the least possible cost, often in the most desirable remaining open space within their jurisdiction. Thus streams were dammed or polluted, ridge lines and scenic highways were fenced in by transmission lines, and wetlands filled, either for new development or garbage dumps.

The rate of development increased with the impact of the population and technological growth following World War II. The accompanying and ever-more-visible loss of clean air, water, and open space, however, led to a counterwave of conservation thinking and discussion, culminating in the latter part of 1969 and early 1970 with a massive shift of public opinion that *demanded* consideration of the environmental aspects of nearly every facet of our lives.

This change of goals, however, finds us with a complex structure of laws, government, and taxes still predicated upon the developmental goals of our earlier national history. From the framing of the Constitution and adoption of the Northwest Ordinance in 1787, we have steadily and carefully constructed a system which not only encourages and assists development, but practically demands it.

Major changes in some of our most basic laws are thus necessary if our legal system is to provide incentives and channels for environmental protection. It is at this point in time—as our legislators and governmental officials look to the challenge of creating new laws to modify the old—that we begin to realize how woefully unprepared we are with knowledge. We aren't *sure* that we should build more secondary sewage plants (a $1 billion federal appropriation in 1970). Perhaps, instead, we should require regional collector systems and deep ocean outfalls, as has been recommended for the San Francisco Bay Area.

We aren't *sure* that we should build a Supersonic Transport ($290 million requested by the President in 1970). We aren't *sure* how we can best stimulate the auto industry to build smog-free or low-emission cars, whether by legislative prohibition (recently passed by one house of the California legislature) or by incentives such as a proposed $35 million federal subsidy to whomever can manufacture a low-emission vehicle.

We aren't *sure* whether continuing population growth poses such a threat that we should liberalize abortion laws in all fifty states as we have in ten. Should we make abortions free to those who can't afford them? Does every child have the right to be wanted?

The answers to these questions, and a hundred like them, can come only from an educated public dialogue, based on a broad understanding of the facts, and of the laws, which presently stimulate rather than inhibit continuing development and pollution. Then, and then only, will legislatures and the Congress respond with the legal changes required.

This book, therefore, can play a major part in developing public knowledge, which must complement already-developed public awareness if appropriate solutions are to be found. It represents, to my way of thinking, the most comprehensive and readable compilation of what is known today about those threats to man's environment which should be understood by every citizen who wants to take part in the great public dialogue of the 70's. I hope it will be available to every public official and student in America, and I commend it to your examination.

PAUL N. MCCLOSKEY
August, 1970
Moose, Wyoming

INTRODUCTION

THE AMERICAN PUBLIC has been told repeatedly that the Environmental Decade has just begun. These are the times in which we are to restore our waterways, clean our air, get rid of the litter and junkyards, build new towns and cities—all for a world in which prosperity and success are to be measured in terms of *quality, not quantity.*

Quite frankly, I'm not sure just what is meant by this. At least I am uncertain as to how seriously it is intended *or is being taken.* While I personally feel that a wholehearted commitment must be made if our children and their children are to face a future that is not truly grim, it is obvious that as yet we are not embarked on a national program of the scope and magnitude of the space effort, so that we can preserve and manage *our own planet* with the same marvelous efficiency with which we have probed the galaxies. The bulldozers seem as busy as ever trying to improve on nature, and the land acquires millions of acres annually of new asphalt and concrete cover. I can still buy pesticides that have been "banned" as health hazards. Just as there is no committee in Congress devoted exclusively to urban affairs, so there is no committee whose only mandate is the drafting of legislation to protect and manage the environment. The committees that draft antipollution laws or consider the adverse effects of technology are also beholden to exploitive interests or are not in a position to do more than educate the public.

It has struck me that the issue of environmental quality has foundered, even up to now, because it is too often assumed that *everyone* is really for it—so why get excited—or that there is little that a citizen can do, short of keeping his yard tidy and not being a litterbug.

As the title indicates, this book is an attempt to show what is being done, what some of the alternatives are, and where you can look for assistance in taking on environmental challenges. It is an attempt *to create a state of mind, namely the understanding that the objectives of quality actually pay off in real benefits* as well as in appearance. My own pessimism as to how we are pursuing environmental objectives thus is not the theme of this book.

The project was suggested by David Dominick, whose duties as commissioner of the Federal Water Quality Administration prevented him from participating. As I understood him, he proposed a book doing three things: 1. defining the problems in basic terms; 2. identifying the main causes; and 3. explaining what all concerned *could or should* be doing about them.

There was no time to lose in putting it together. Rather than spend a year doing fresh reporting, I felt it would be far more productive to make use of notes taken as a journalist during the past two years, supplemented by research into what was presently understood and recommended concerning environmental imperatives.

Thus thanks are very much in order to the many who assisted me in compiling data and to those who suggested ways of handling all this material. Sidney Howe, president of the Conservation Foundation, has been particularly patient and helpful, contributing his own thoughts and then putting me in touch with the specialists in his very impressive organization, some of whom are cited in the book. Shirley Briggs, executive director of the Rachel Carson Trust for the Living Environment, Inc., was kind to allow me to use her pesticides guide.

My sister, Sally Saltonstall, wrote letters to some forty congressmen and senators and some eighty conservationists and other "participants," including businessmen, throughout the country. These persons were all asked for their ideas, copies of proposals they had made, and any material they could contribute. The response was gratifying to the extent that it would be impossible to cite all those who responded and influenced my thinking. Many are mentioned in the book.

Sorting out all this material as well as a huge backlog of clippings, notes, and technical reports was the task of my wife, Emmy, and ten-year-old son, Dicky. Jake Page, my editor, never failed to come up with a good suggestion for organizing each chapter. And finally, to get it all finished in time, my sister and her friends Annie Cunningham and Judee Metcalfe worked after hours on Capitol Hill to produce 327 pages of manuscript in a few nights. This, I hope, resulted in a book that is fairly up to date in a field that is continually exploding with new ideas.

GREAT FALLS, VA.
JULY 20, 1970

1 ⚜ WATER

WATER IS THE RESOURCE that most *visibly* has affected our way of life, our settlement patterns, our prosperity and—not least—our health. And because it touches our living and working habits in so many ways, it is understandable that no other resource has been more abused. It is bad enough that we have polluted the water that we drink. But impure water is only part of the problem. At the outset, in a chapter about water, it cannot be emphasized enough that much more is at stake than making water clean. It is inefficient use of water—wasting it, failing to control it, or controlling it the wrong way—that imposes unbearable environmental strains. When water imbalances are created, by industrial squandering, municipal misappropriation, and by allowing the resource to run away and erode the land, or when brilliant engineering systems mislead people into thinking that there is plenty of water available, pollution is the result.

This may seem obvious, but we so often seem to overlook the ultimate effect of biological disruptions. For example, the community that has at last obtained sewage treatment to return water clean to lakes, streams, and the ground, may have a severe pollution problem because of erosion and runoff from residential developments or a new shopping center.

Or consider how misappropriation of water has led to a catastrophe that keeps escalating in southern California, a region that contains more than half the state's population but receives only 1.5 percent of California's rainfall. The sparsely settled northwest coastal area gets 38 percent of the state's natural water supply. Elaborate, costly systems of dams, canals, pumps, and pipelines have been devised to quench the southern half's thirst. Explosive growth around Los Angeles is thus encouraged and weird and unexpected things happen—climatic

inversions that cause eye-burning, health-impairing, crop-kill-
ing pollution; winter torrents that cause lethal mudslides and
summer dry spells that bring on rampaging fires in the chap-
paral-covered hillsides to finish the havoc the rains began.
Los Angeles is an artificially created environment because it
steals the one resource it lacks—water.

On the way south, the water is denied to river systems that
used to feed rich valleys, severely limiting the growth of settlements in these regions. Instead of moving toward the sea naturally, the water is rationed through irrigation canals to encourage more artificial communities. Now scientists have shown
that the $3 billion California Water plan, all but 25 percent
built, is an ecological disaster. For example, one of its key
links will upset the biological balance of the Sacramento River
delta and impair the water quality of the San Francisco Bay
into which it feeds. This link is a canal to divert water from
the river to the San Joaquin Valley fifty miles south. Ecologists
were scoffed at when they said years ago that cutting down the
Sacramento's natural flow would critically change the delta's
salinity and would allow more sewage wastes to accumulate in
the bay. However, these contentions were supported in a report
issued in the summer of 1970 by the U.S. Geological Survey.
(See Bibliography.) One could catalogue forever the chain-like
effects of California's water planning. It is not wrong to build
desert cities with water from somewhere else, but matters easily
get out of balance. The tragedy is that cities like Los Angeles
and Phoenix keep right on growing, becoming the foci of desert
megalopolises. These are dry and sunny places—good for what
ails you—but their price is high when measured by the denial
of water benefits elsewhere and by the environmental disturbances induced by mismanagement of the resource. "We insist
that water must be shipped to places where people and industry
have located," writes ecologist Raymond F. Dasmann in his
exquisite treatment of California's dilemma. (See Bibliogra-

phy.) He adds that "We could equally well insist that people and industry should locate in the areas where water is available."

And that is the point that must be kept in mind. Water is a critical element of land use and community planning. But too often is this fact overlooked when people decide where they want to work and live and when industries decide where they want to manufacture and trade. One can only conclude—and this point will be discussed again—that a price must be affixed to water that measures its worth in environmental terms. Until that happens, we will continue to be profligate in our use of it. Warning messages have been flashed repeatedly in Los Angeles. There will be further consequences there and elsewhere. "If our ability to manage water falls short," Dasmann warns, "the entire framework of civilized life is threatened."

As we enter the decade of the seventies, the framework has never been shakier. The great rivers that opened up the New World—the Hudson, Connecticut, Delaware, and Ohio, the Mississippi, Missouri, Columbia, and Colorado—are dead or dying from the effects of domestic and industrial effluent, silt, and thermal discharges. Some, such as the Cuyahoga, the Buffalo, the Grand Calumet, and the Mahoning, are so slick with oil and chemicals they are seriously considered to be fire hazards. All but the very largest lakes, and even they are beginning to show wear, are contaminated by chemicals or pesticides and are literally suffocating from an alteration of the oxygen balance known as eutrophication. The estuaries, where fresh water meets tidal salt water, are choked by sediment and garbage and have lost their full value as marine habitats and inshore fisheries. (See Chapter 8.) Marshes and ponds over-enriched by phosphorus and nitrogen have also become bogs before their time from accelerated eutrophication or they have been filled in for land and agricultural development often directed by a federal representative who should know better. It is unsettling to discover that even deep groundwater supplies contain traces of health-endangering elements. Recently, for example, the Snake River Plain aquifer in Idaho was found to be contaminated by radioactive wastes from the Arco-Idaho Nuclear Test facility.

If you cannot apply at least some degree of these conditions to your local body of water or drinking supply, then you are lucky indeed. Even then the chances are good that someone upstream, or at the upper end of your watershed, will eventually poison your water.

FRAGILE RELATIONSHIPS

The water or hydrological cycle is a simple process. Water arrives from the atmosphere in a rainfall. And the United States is rain rich, with an annual average of thirty inches. After evaporation, this amounts to 1,794 billion gallons a day. (Alaska accounts for a whopping 580 billion gallons of this, and Hawaii 13 billion gallons.) Perhaps a third of the rain percolates through the soil and into the water table. The rest evaporates back into the atmosphere after transpiring through vegetation or running into streams and thence to the oceans.

Not so simple is the way water affects living things and plant organisms. At each step in its use, a fragile relationship must be maintained. Slight changes in flow, in temperature, in ratios of oxygen and hydrogen, in color and turbidity, in nutrient and mineral composition can be consequential. For example, a warming of merely five degrees will kill trout and salmon. These fish not only provide food (and sport) but exert a healthy influence on their environments. Traces of chemicals will magnify as they are passed along the aquatic life chain to the point at which fish and birds die or become infertile. In the end, even humans are affected. Fertilizers and other nutrients, such as phosphorus from detergents, will inaugurate the eutrophication death cycle by stealing massive amounts of oxygen. These slight disruptions add up. On top of them come the big blatant kills. The Federal Water Quality Administration publishes an annual count of fish dead from identifiable sources of pollution. One should assume that when the fish die, so do the tiny, uncounted aquatic organisms. In 1968, over fifteen million fish were victims, 31 percent more than in 1967. The largest slaughter resulted from carelessness on the Allegheny River at Bruin, Pa., where a petroleum refinery's waste lagoon overflowed into a pond. Its banks then burst and chemicals poured into the river in a wave of suds six feet high. In Mobile, Ala., an overloaded sewage treatment plant had to pump excess effluent into the Dog River, so lowering that water's oxygen content that more than a million fish suffocated. In fact, inadequate city sewer systems were the main sources of trouble as they were responsible nationally for seven million fish kills in 122 reported incidents. Next came industrial wastes, then "transport accidents" such as a broken pipeline or a train wreck. Some of the causes were ridiculously unnecessary: cleaning acid that drained from a swimming pool into a creek; spilled sheep disinfectant; methyl and ethyl parathion that drifted from a cropduster plane, a steelmill banking its blast furnace, and a caustic cleaner that was flushed into a storm sewer instead of into a sanitary outlet. The dictionary says that to pollute is "to make physically impure or unclean." Well, pollution also kills.

POLLUTION'S MANY FORMS

What are the insidious and complex forms of water pollution? What are tolerable levels of human practices known to

have harmful effects on our water systems? How can we develop
an early warning system? What water quality standards have
already been set and how does one benefit? What new and
broad approaches to water resource protection and management
must be developed if we are going to maintain a clean and
sufficient supply of fresh water into the future?

Pollution is always popping up in new places, so it doesn't
help much to describe it just as an aftereffect. It is worth
remembering that no sooner had the government acted against
DDT in 1969, than 2,4,5-T and mercury caused trouble. Mercury is a waste product of many chemical manufacturing processes, of pulp and paper plants and other industries, and is
used to coat agricultural seeds. In minute doses, mercury is extremely dangerous to humans, killing brain and nerve cells and
affecting chromosomes. Like DDT, it is persistent, and already
has built up to dangerous levels in fish and wildlife. In
mid-1970, fishermen were urged not to eat the fish they
caught in the St. Lawrence, Oswego, and Niagara Rivers, and
Lakes Erie, Ontario, and Champlain. A federal task force
turned up alarming evidence of mercury contamination in
Alabama, Delaware, Georgia, Kentucky, Louisiana, Tennessee,
Texas, Maine, Michigan, New York, New Jersey, Ohio, Pennsylvania, Vermont, Washington, West Virginia and Wisconsin.
Dangerous traces of mercury were found in every TVA reservoir
in the Tennessee River system. Health officials can control future
wastes but have no idea what to do about the mercury already
present, short of telling people to reduce their water intake or
stop drinking it entirely. The Food and Drug Administration
has set a tolerance of .5 parts per million for mercury in fish and
animal tissues but has been considering a lower level. Ten percent of the mercury you consume goes to your brain, where the
cell damage it causes may not be noticeable for years. Who knows
what, when, and where the next villain will be. It is more
useful in developing a feel for prevention to understand the
activities that generate pollutants, in hopes that new codes of

use and respect may be born and that this nation will, like Sweden, adopt the rule that "If you don't know what its side-effects will be, don't use it." The current rule in the U.S. today seems to be, "Don't worry about it until somebody or somebody's food gets poisoned."

Domestic Sewage and Eutrophication

Home waste is the most troublesome source of pollution. While industrial effluents are major contaminants, generally they can be pinpointed—if not defined—and remedied legislatively or legally if citizens are sufficiently aroused. But attacking the domestic load of sewage is another matter because it means curbing people's habits of water consumption for cooking, cleaning, flushing toilets, washing cars, and so forth. The problem is complicated further because water experts are forever mystified by the interactions of pollutants under certain water conditions and this confusion often gives those who appear to be the source of trouble an excuse. Take the eutrophication crisis.

There is hardly a stretch of water that does not suffer to some degree from excess growth of algae and weeds because our wastes are over-fertilizing aquatic growth. Attacked by bacteria in the process of decomposition, these wastes use up oxygen and then release more nutrients that in turn nourish the underwater crops. The nutrient components such as phosphorus and nitrogen boost the process considerably. So-called algal blooms use up more and more oxygen in a vicious cycle that ends up with a lake becoming a stinking jungle of weeds, its fish dead from oxygen depletion. Florida's Lake Apopka, once full of bass, is now being drained, at a cost to the state of $140,000, so that the effects of eutrophication can be baked away and the lake refilled to be navigable and fish-productive once more. It will take over a year and is not considered a

solution for the nation's larger lakes. It was too late to try
anything else to save Apopka.

How do you prevent eutrophication? Most scientists believe
that phosphate-based detergents contribute up to 60 percent of
the phosphorus that causes this condition. The soapmakers
argue that other sources in industry and agriculture are equally
to blame and that not enough is understood about the role of
carbon and nitrogen in contributing to the algae growth. The
experts confess that, true, there are plenty of unknowns about
eutrophication, but the hard evidence at hand does indeed indicate that a substitute water-softener for phosphates would
help a lot. At hearings in December, 1969, concerning this
issue, Wisconsin Congressman Henry Reuss quite rightly refused to bide with the line that phosphate detergents could not
be restricted until replacements were found effective. "Isn't this
a little bit as if you heard that the headmistress of a school was
feeding her kids arsenic but you didn't dare ask her to stop for
fear she'd feed them something worse?" he asked.

Tests have been promising on a phosphate substitute, ni-
trilotriacetic acid. However, so far it is more expensive and not
stable in warm, moist climates. (It tends to cake.) Even so, it
is estimated that acceptable NTA would cost the ten million
U.S. (excluding Canadian) citizens in the Lake Erie basin only
an additional four million dollars annually, or forty cents per
person, a small price to pay to clean up the nation's most
celebrated cesspool. Tertiary waste treatment plants to remove
entirely the phosphorus from sewage would cost $230 million
for the Lake Erie basin, according to federal estimates. So,
clearly, a new detergent is desirable.

Sears and Roebuck was in fact set to market a phosphorus-
free detergent in September, 1970, under its De Soto Co. label.
Preliminary tests by the Federal Water Quality Administration
were promising, although the biologists were mystified as to how
Sears could come up with a detergent without even a substitute
base for a phosphate formulation. They found the Sears product

cleaned efficiently even if they had not completely ascertained its components and their ecological side-effects.

While industry and government try to reach accord on new detergents or if the Sears and similar products fail the environmental test, what can you do about it? For one thing, you can buy a dish cleaner or laundry detergent that has a low percentage of phosphates, or you can use straight soap and soda. (See Chapter 6.) Action groups across the nation have put out detergent lists. And the FWQA has also published its laboratory analysis of phosphate levels, measuring the phosphate most widely used in detergent formulations, chemically described as sodium tripolyphosphate (STPP).

Detergent Phosphate Count

TYPE OF MATERIAL	PRODUCT	PERCENTAGE PHOSPHATES AS STPP
Pre-soaks	Biz	73.9
	Enzyme Brion	71.4
	Amway Trizyme	71.2
	Axion	63.2
Laundry Detergents	Blue Rain Drops	63.2
	Salvo	56.6
	Tide	49.8
	Amway SA-8	49.3
	Coldwater Surf	48.2
	Drive	47.4
	Oxydol	46.6
	Bold	45.4
	Cold Water All (powder)	45.4
	Ajax Laundry	44.6
	Cold Power	44.6
	Punch	44.2

	Dreft	41.9
	Rinso with Chlorine Bleach	41.0
	Gain	39.5
	Duz	38.3
	Bestline B-7	38.0
	Bonus	37.5
	Breeze	37.2
	Cheer	36.3
	Fab	34.8
	White King (with Borax)	34.7
	Royalite	21.7
	Instant Fel Soap	16.6
	Wisk (liquid)	14.2
	Par Plus	4.3
	Addit (liquid)	2.2
	Ivory Liquid	1.9
	Lux Liquid	1.9
	White King Soap	less than 1.0
	Cold Water All (liquid)	less than 1.0
Automatic Dishwasher Detergents	Amway	60.0
	Cascade	54.5
	All	54.0
	Calgonite	49.4
	Electrosol	34.8
Household Cleaners	Ajax all Purpose	28.5
	Mr. Clean	27.0
	Whistle	3.1
	Pinesol	less than 1.0
Miscellaneous	Snowy Bleach	36.4
	Borateem	less than 1.0
	Downy	less than 1.0
	Amway Dish Drops	less than 1.0

* It should be noted that while phosphate contents are reported as percentage STPP, not all products contain STPP.

You can also help to exert pressure for a new soap base substitute that will produce less phosphorus by writing letters to soap manufacturers and, more important, to your congressmen, telling them to support proposed legislation to ban or restrict phosphate-based detergents. Finally, you should explore the possibilities in your particular region for reusing the phosphorus in fertilizer or some other way. More will be said about this, but already several pilot projects have been successful in removing phosphorus from sewage and making it a benefit.

Waste Treatment

Some thirteen thousand communities, containing 68 percent of the nation's population, have sewer systems. But in fact, only 40 percent are adequate, and a thousand sanitary systems are outgrown annually.

With the problems of phosphorus still in mind, this is a good time to summarize what happens to most domestic sewage and a good deal of industrial waste.

To begin with, treatment processes are no more than a speeding up of natural processes of self-purification. That is why many harmful, nondegradable or extremely persistent and toxic chemicals are so hazardous. They pass through treatment unaffected. The most important concern of sewer plants is to reduce the waste's demand for dissolved oxygen by performing bacteriological decomposition before the waste is returned to the waterways.

There are three stages of waste treatment—primary, secondary, and tertiary. Primary treatment amounts to no more than a brief holding action. Sewage is filtered through screens to rid it of debris. It passes into a grit chamber to allow sand and gravel components to settle. A sedimentation tank removes more particles, which become sludge. Then the sewage is chlorinated to kill bacteria and remove odors before it is discharged

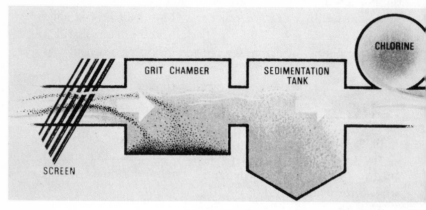

PRIMARY TREATMENT

into a waterway still containing oxygen-demanding organic matter. Secondary treatment does a more efficient job by adding several stages, but the discharge is certainly not cleansed of major nutrient components, its full demand for oxygen, or the impervious chemicals. Either a trickling filter of stones or, more likely, an activated sludge process is used. The latter consists of an aeration tank in which sewage is combined with bacteria-loaded sludge and air so that organic wastes will be decomposed further. From the aeration tank the sewage proceeds through a sedimentation tank where solid particles are removed, and then it is chlorinated. The sludge is reprocessed until it piles up and has to be disposed of somewhere. Getting rid of sludge is a major problem. For example, Chicago produces one-thousand tons of solids from sludge daily which cost over $18 million a year to dispose of, mainly as fertilizer and landfill.

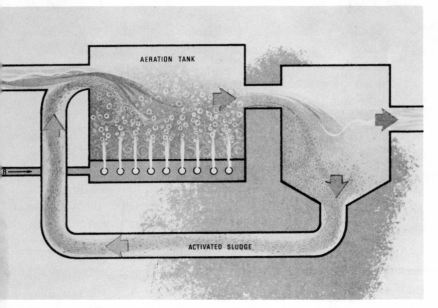

SECONDARY TREATMENT

Tertiary treatment refers to various advanced processes that produce drinkable water. All are expensive and none has been proven applicable nationwide. It is much more likely that a combination of solutions will be desirable instead of universal tertiary treatment. Where potable water is in short supply, advanced technology will be necessary. One technique that appears promising has been tested at the Blue Plains plant on the Potomac River. It combines conventional biological treatment with chemical applications and the use of pure oxygen (instead of air) to remove 100 percent of the biological impurities, 96 percent of the phosphates, and 85 percent of the nitrogen in waste water. But such a plant to handle a city load in excess of 300 million gallons a day will cost well over 300 million dollars at today's money value just to install.

In areas where the water supply is sufficient to allow longer recycling periods, there are infinite possibilities for handling waste. One method tried by several localities is to send waste water through irrigation systems after basic treatment. Before it makes its way back into rivers and lakes as groundwater, the waste is filtered free of its nutrients and has helped nourish crops and forests. Chicago, for example, sends some of its waste water by rail to Arcola, Illinois, where it is sprayed on cornfields. Aided by a federal grant the city and county of Muskegon, Michigan, have purchased large tracts of land just outside the city which will receive loads of sewage nutrients to fertilize fields of corn and other crops. Certainly there will be more revelations concerning the use of waste elements that reduces the need for, or the scale of, advanced waste treatment. Water quality enforcement and other means—legislative as well as economic—are called for to bring about marketable waste by-products while enhancing the nation's waterways.

Another alternative in the handling of sewage is combined treatment, whereby industrial and municipal wastes flow through the same plant. Again this is not the universal remedy that some industries, looking for an easy ride, say it is. Combined treatment can be beneficial, as when industrial waste neutralizes the dissolved oxygen demand of domestic effluent and thus introduces a new stage of sewage decomposition. However, frequently factory wastes are unique and thus—from the public standpoint—could be far more effectively controlled by recycling techniques. Moreover, industries generally do not put a high value on the quality of the water they *take in*. If it is used in a manufacturing process it generally requires specialized treatment unless it is extraordinarily clean to begin with. Otherwise, they use water mainly to provide power through the steam cycle. For years, conservationists have said that industries should be forced to put their water intakes *below* their outfall pipes, so that they would have no choice but to clean their wastes. This is a simple—although ideal—solution that does not seem likely because of politics, but a number of incentives *and* penalities are being considered, as will be noted later in this chapter, that will help to make factories responsible for their own waste treatment.

Industrial Effluent

Well over 300,000 U.S. factories use water. It is estimated that they discharge about three times as much waste as domestic sewer systems. And industrial effluent is obviously the most toxic. A 1970 FWQA report, "The Cost of Clean Water and Its Economic Impact," lists some fifty-one identifiable troublesome industrial agents from nondegradable chemicals to materials demanding an inordinate amount of oxygen in the water. The major contributors are pulp and paper industries, petroleum refineries, organic chemical manufacturers, blast furnaces, and steel producers. Unfortunately, there has not been a federal inventory of these wastes while their volume is growing several times as fast as sanitary sewage. And highly specialized industrial toxins, dangerous acids, radioactive substances, and mineral particles combine and interact in upsetting ways underwater. Their synergistic effects are barely understood by water biologists. Officials increasingly are worried about the effects of chemical and metal traces never before considered particularly dangerous, such as copper and zinc. Recently, minute amounts of copper alarmed biologists by harming clams, turning the mollusks' insides green. The 1970 federal report estimated that industries would have to invest $3.3 billion or more over five years in waste treatment facilities. Of course such estimates have assumed that various manufacturing processes would use water the way they do now, but most scientists feel that these methods of water consumption are inefficient and wasteful.

Federal officials feel that if industrial water users would bear in mind growing demands on the water supply as well as increasingly tough water quality standards, the businessmen would conclude that in-house treatment facilities will pay off in profits. (Present U.S. daily use of 269.6 billion gallons is expected nearly to quadruple by the year 2000.) Moreover, it would appear obvious that recycling systems add efficiency to a manufacturing process because they enable an industry to recapture products from the treatment sludge. Such observations however, have not been adequately analyzed except in a few instances, and thus the evidence is only preliminary. Some pulp and paper companies, for example, have discovered that recycling will pay off in added efficiency if the system is coupled with plant overhaul that was due anyway. Kaiser Steel has been recycling water in its Fontana, Calif., plant for some time, using a fraction of the volume of water needed by a conventional mill. But this innovation was forced by a limited water supply. "Industry in general," says David Dominick, aggressive young commissioner of the FWQA, "has been extremely short-sighted in reacting to tougher antipollution legislation. But we are fast coming to the point of stating that legislation will ultimately force reuse mechanisms to be built into the production system." Until industry acts as if that climate already existed, citizens should help to create it by criticizing the companies in which they hold stock or by urging local and state officials not to allow new plant construction that doesn't incorporate the latest possible waste treatment devices.

A century ago, an English pollution commission had already concluded that antipollution legislation should be so tough that factories would have to clean up their production processes. The English took a comprehensive and remarkably sophisticated look at what the industrial revolution had wrought on the island's renowned rivers and streams. They analyzed the effects of wastes dumped from lead, tin, coal, manganese, and other mining operations as well as from industries using or burning these products. Addressing them all, the commission ruled that "The remedy required for the present reckless discharge of solid refuse from mines into the neighboring river channels is simply the enactment of an adequate penalty. It ought to be at once forbidden." Thus, in a moral pronouncement, the English investigators ruled that a civilized, technologically superior nation had no business tolerating pollution. Clean up the mess first and work out the sharing of costs later was their edict. Unfortunately, it didn't go into full effect, although some clean-up began and in recent years has gained a good lead on U.S. efforts.

Not until 1899 did a U.S. law go on the books saying "Thou shalt not pollute." But this law was virtually ignored until Wisconsin Congressman Reuss uncovered it in 1970 and urged Americans to take action. His office put out a pamphlet explaining the law and telling how citizens could reach the right authorities. In sum, the old measure stated that any person who discharged refuse into a navigable waterway was committing a federal crime and could be fined up to $2,500 or jailed a year. Better yet, after reporting a violation to a U.S. district attorney, the informer was entitled to half the resulting fine as his reward. (See Appendix I, the contents of Reuss explanation.) Unfortunately, as will be noted in Chapter 10, the U.S. Department of Justice was reluctant to prosecute complaints citing the Refuse Act even though it remains the most stringent anti-pollution law in the Federal book.

Industrial pollution control, the possibilities of recycling, and the assessment of costs and benefits are the main ingredients of a waste crisis that threatens to get out of hand, if it hasn't already. It suffices to say here that strong citizen-consumer advocates would never let a business pass all of the bill

along to the buyer, particularly if the company had been polluting the air and water flagrantly and at the expense of the
consumer's health.

Effects on Drinking Water

The deteriorating quality of our drinking water, not to mention ill effects caused by multiple chlorination during droughts
when reservoirs are too low to filter, naturally makes solutions
to water pollution all the more urgent. Cities across the country
are drawing their water from lake, river, or ground sources into
which every known pollutant trickles. It is impossible to trace
or even inventory all the chemical components of drinking
water supplies, so water treatment is handicapped in the very
beginning. Long-term, low-level exposure to such contaminants
poses a health hazard. A 1969 report by the American Chemical
Society (see Bibliography) notes that nitrates are known to
cause methemoglobinemia and death to infants if present in
drinking water above forty-five parts per million. Little is
known about the characteristics of and ways of treating water-
borne viruses. While chlorine has been used since 1915 to
disinfect water by killing bacteria and viruses, there is alarming
new evidence that it too poses a health threat if overused. Certainly chlorine by itself is known to be a highly poisonous gas.
Federal health officials were disturbed by a recent survey that
showed chlorine was being handled across the country with
extreme carelessness and disregard for the consequences of over-
application, and Stanford medical researcher David Peter Sachs
reports that genetic studies show that multiple chlorination
treatments may damage the DNA, the essential life substance,
in human and other mammalian cells. He even suggests that
high chlorine levels in drinking water might produce cancer of
the bladder. Public Health Service surveys regularly detect that

the nation's municipal drinking supplies are contaminated by trace chemicals and minerals, do not pass bacteriological standards, or are over-chlorinated. There is enough disturbing evidence to urge that citizens check out their drinking supplies—where the water comes from and how it is treated—by inquiring of local and state public health and water district officials.

While domestic sewage and basic industrial consumption are the main causes of water pollution, there are other actions that should be understood. In these cases the blame is widely shared.

Thermal Pollution

Such is the term used to describe an addition of heat that impairs water quality, invariably by speeding up biological processes, reducing the dissolved oxygen supply, nurturing aquatic plant growth, and reducing a water body's capacity for further use in cooling. These impairments obviously harm marine life. Thermal pollution is caused by heated waste water that has passed through the condensing stage of the steam cycle, usually in a steam-powered electricity plant. Since the nation's appetite for power doubles by the decade, the volume of waste water from utilities has been rising dramatically.

Nuclear power steam plants pose the greatest thermal threat because they are generally three times as large as conventional fossil fuel (coal or oil) steam plants, and the nuclear-steam cycle, while it is cheaper to run, is far less efficient. It uses up to fifty percent more condenser water. Rises of temperature in cooling water returning to a stream range from ten to twenty degrees, sometimes go as high as thirty degrees, and average thirteen degrees. This is serious when you stop to consider that at the current rate of growth the power industry will within a decade demand one-fifth of the total freshwater runoff in the U.S.

It has been proven that the slightest water temperature

changes adversely affect fish and it is feared that standards now
being set do not adequately protect tiny plankton and microscopic, sensitive organisms so vital in the aquatic food chain.
It is known that creatures accustomed to a seventy degree environment will die in ninety degree water. Spawning cycles are
upset by thermal infusions that set up heat barriers through
which fish cannot pass and, at a million gallons a minute (a
typical discharge), have an effect far downstream. In 1968, a
federal task force (see Bibliography) recommended thermal
guidelines for varying conditions. These ranged from ninety-
three degrees for tolerant catfish, gar, certain bass, and shad to
forty-eight degrees for spawning lake trout, walleye, and northern pike. But if you are interested in determining criteria for
your waterway you should not be governed strictly by published
guidelines. Rather you should meticulously investigate water
characteristics and the needs of indigenous plant and fish life.
Check with the regional FWQA agents or seek advice from
university biologists. It is a sad reflection on the local political
climate and behavior of large economic interests that not one
state has yet given the highest possible classification to all its
waterways or set criteria that are tough enough with respect to
all water uses, although some states have set standards exceeding
federal criteria for particular uses. An example is Minnesota's
criteria for nuclear plant discharges.

What are the remedies to thermal pollution? The 1970 Water
Quality Act provides that the Atomic Energy Commission cannot issue a nuclear power plant license until the applicant has
satisfied state water quality officials' criteria to prevent thermal
pollution. But few states have set tough guidelines, and not
nearly enough has been done at any level of government to
coordinate energy systems and plan for future requirements,
notably plant siting plans and how they will affect the environment. Revolutionizing the steam cycle or finding a use for
the waste heat would help. Cooling towers and ponds are

recommended, although some scientists contend that stack vapors could create a fog problem and ponds require land that is generally scarce or unavailable. One point has been made by the FWQA: If the utilities prevented thermal pollution with current and approved technology it would add only one percent or less to the monthly electric bill.

Perhaps the most alarming and certainly the least understood aspect of nuclear power plants is not their potential for thermal effects but instead their capacity for long-term, cumulative radiological contamination. Critics of the Atomic Energy Commission have raised valid questions concerning current radiation health standards. A federal committee is working to develop new radiological criteria, but an independent investigation would be more effective. The storage of nuclear wastes, underground nuclear testing, and other "peaceful" uses of the atom are threats to stored and flowing water. No one knows how the effects of radiation traces will accumulate and magnify as they move downstream. And the organization of government does not presently give adequate opportunity of advocacy or sufficient legal and scientific recourse to those who are critical of the way nuclear energy is developed.

Oil and Hazardous Substances

While oil spills have been more celebrated over the past three years (since the breakup of the Torrey Canyon) as oceanic, saltwater disasters, oil leaks, sloppy handling of oil, and lax methods of disposal have increasingly been menacing lake and river habitats. (Oil as it affects estuaries and the inshore fisheries will be discussed in Chapter 8.) Gas service stations alone annually dispose of 350 million gallons of used oil. There is no need for it, because reprocessing oil that has been used in crankcases or in manufacturing is feasible and economical.

Moreover, some reprocessed oil is already sold, but Federal Trade Commission labeling restrictions have undermined the market for such products by making them seem inferior to new refined oil. Thus millions of gallons of oil daily get flushed down the drain. These wastes are apt to bypass sewer treatment plants, or if they end up with industrial and domestic sewage they can cause havoc. In Nashville, fumes from waste oil and gasoline exploded and put a sewer plant out of operation for several months. In any event, the present waste treatment operations do not break down the oil (or gas) enough to make much difference. Recycling is the only answer.

Trains, tanker trucks, and barges are moving all over the country with extremely volatile cargoes, flammable liquids, explosive gases, and killers such as cyanide. It is doubtful that this can be halted entirely, but experience demands many more restrictions and handling standards. For there are presently ten thousand spills a year of such pollutants on the nation's navigable waters.

Agricultural Sources

Feeding the nation is not without unpleasant side-effects. Farm livestock and poultry excrement, and wastes not used in by-products add up to three times the volume of human sewage. The ratio increases as we are caught up in the agricultural revolution. (It would be a good revolution if it did not take for granted a growing number of people, and so much of its production were not destroyed in the U.S. to keep farm prices high.) Livestock are no longer fattened on the range where their manure is naturally dispersed in helpful amounts. The feedlots are concentrated near transportation centers, and the great stockyards where the cattle are slaughtered are in big cities like Chicago. It is not an exaggeration to say that cattle manure has become a major urban problem. It is a powerful nutrient. Animal wastes in general clog or complicate sewage processes, add bad tastes and odors to water, demand high amounts of dissolved oxygen, and can produce harmful bacteria. Feedlot pollution is critical now but it is going to get worse. It is estimated that our current population growth, coupled with food consumption, demands each year another 430,000 beef cattle, 61,250 dairy cattle, and 1,082 hogs. No one has yet coped with this crisis and the only obvious cures would seem to be improved handling of the wastes, pressuring feeders and processers to install abatement devices, or finding new, and increasing old, applications for wastes. A promising precedent has been constructed by the Midwest Feedlot Co. in Cedar Creek, Nebraska. Previously the company was pouring manure into the headwaters of Cedar Creek, despite the use of lagoons to hold animal wastes. Federal abatement threats forced the feedlot managers to build a huge earthen dam to capture *all* water runoff from the hillsides where the cattle feed in pens. Above the dam, rock weirs were constructed to capture solid particles which could be hauled away for fertilizer. The liquid runoff then was pumped into spray irrigation lines to enrich neighboring cornfields. "This isn't a waste treatment problem," says FWQA's Dominick. "It's a waste management problem." And without question, local, state, and federal farm consultants and planners have failed to consider good water management in encouraging forms of land use and methods of farming. The consequences of overusing water are felt as irrigation systems fill up with nitrates and soil salts that are in turn carried into rivers and lakes. The Colorado River has become so saline that U.S. management of its water is under attack by the Mexican government whose farmers have been thus denied full benefit of the river downstream. Better lining of irrigation canals and desalinization are possible answers. But more to the point would be sensible use of the Colorado.

Mine Drainage

Careless past practices and continuing sloppiness in mines have degraded millions of acres of land and thousands of miles of streams in the U.S. In the small Kanawha-New River Systems in West Virginia, for example, some sixty-seven million gallons of mineral acids appear each day, much of this volume from abandoned coal mines whose legal ownership is virtually impossible to pin down. These acids are formed by air and water reacting with sulfur-bearing minerals to make sulfuric acid and iron compounds. More than four million tons of sulfuric acid annually attack U.S. streams and in Appalachia over ten thousand miles of river are polluted by coal mine drainage. However, throughout the U.S., phosphate, clay, iron, copper, aluminum, and other mine operations cause pollution.

And acid contamination is only part of the problem. Massive erosion from leveled and gouged land is also serious. It is estimated that about two million acres of unreclaimed stripmines may be losing over ninety million tons of sediment a year. Apart from being a land crisis (see Chapter 7), this erosion chokes stream habitats, causes flooding by carving new runoff routes or clogging up old stream beds, and changes the turbidity of rivers and lakes adversely.

The solution to mine drainage and sediment erosion is far tougher legislation—state and federal—to force renovation, re-grading, revegetation, and sealing of openings, where it is necessary, in mines under active or known ownership. According to federal studies, it would cost upwards of seven billion dollars to clean up the mine mess. So many mines are in economically depressed areas and are thus important regional assets that it will be politically tough to exert pressure for a clean-up without a strong push by state and federal governments. And because so many mines have been abandoned, federal subsidies will be necessary.

Erosion in General

Every river system in the U.S. provides a case study for examining the effects of land management on water systems. The small brooks and tributaries, marshes, and mere rivulets of water have been filled in for housing or commercial development at a reckless pace. Often they were diverted or confined in culverts, thus limiting their drainage value and creating dangerous flood conditions in rainy periods. The water that was blocked had to run off somewhere, and it did, usually carrying loads of earth into a storm sewer or carving new and destructive paths. Then construction along these water-courses stripped the adjacent lands of the vegetative cover necessary to regulate the flow of runoff. Again, loads of silt materialized. It is now figured that the amount of sediment piling up in the Potomac River each year—2.5 million tons—if not dredged, will fill up the river estuary with solid dirt all the way from Chain Bridge, just above Washington, to Fort Foote, a mile downstream, in only fifty years.

The more than a billion tons of sediment estimated to be pouring into U.S. waterways each year are seven hundred times the volume of sewage, and these so-called "suspended solids" increase the cost of water treatment, spoil water for recreation, and add still more nutrients. Fish react critically to water clouded up by sediments, and power turbines, pumps, and irrigation systems all suffer. Unfortunately for those making a case against erosion, the complete costs of this subtle and complex degradation have not been computed. But enough is known about individual effects to describe sedimentation and erosion as extremely critical factors in the war on water pollution. Of course, the more we pave and cover the land with roads and buildings the heavier will be the flow of sediment, since rain deposits are thus deprived of natural sponges that slow down the pace of water drainage and runoff. Consider that a new shopping center of twenty-five acres will cover nineteen acres with roof or black top, making a smooth flat surface to carry the rain faster. The effects of that runoff are seldom considered beyond the borders of the shopping center. As long as the water is sent down a storm sewer, it appears no harm has been done. But at some point in its journey that water is going to help cause erosion simply by adding to the overflow of some body of water or by overloading a waste treatment plant and forcing untreated sewage to go into a river or lake.

Government agencies, notably the Army Corps of Engineers, do not help the situation. (The Corps' deeds will be described again in both Chapters 8 and 10.) An example is the practice of "channelization" under Public Law 566. This is justified as stream improvement, but often the deepening, straightening, and clearing of a river so speeds up and adds to water flow that dirt is washed away from the banks. There is an estimated 300,000 miles of serious stream bank erosion in the U.S., much of it caused by such activity. It does not help either when the Department of Agriculture urges farmers to drain their wetlands into these "improved" channels. The evidence at hand shows that the benefits from controlling erosion eventually outweigh gains from developing land or increasing its productivity by draining it.

In the upper Midwest, where the land is dotted with ponds called prairie potholes, the farmers have begun to realize that seasonal floods are caused not just by a heavy winter snow pack and a sudden spring melting period. A crucial element has been increased drainage of wetlands and upstream improvements that multiply the effect of the spring runoff.

And finally there is hardly a city in the U.S. today that does not receive serious flooding during a thunderstorm. Whole residential sections in many cities regularly pump out their lower levels after a big rain. And yet city fathers seem too often impervious to the moral of such miseries and go right on allowing construction in floodplains and land development that will bring on more water.

Estimates of the costs of preventing or controlling erosion fluctuate so wildly they are useless. Damage done is expensive to repair. But evidence indicates that the traditional rules of landscaping—contouring, terracing, exposing small areas of raw earth for short periods—can be applied to hold sediment at a cost of around $1,000 both on a mile of highway construction and on the excavation required on the average urban building project. Controlling erosion is still mainly an individual act that adds up to no more than common sense on the spot and a sense of responsibility for magnifying side-effects.

If construction activity threatens your water environment with sediment runoff, you should speedily state the case to city, county, or state water quality officials; even find out if the nearest Department of Agriculture agent, who is apt to be a soil conservation specialist, has an interest. Many localities have set enlightened zoning standards to curb erosion. These range from provisions against building on steep slopes to density restrictions along waterways. More will be said about these examples in Chapter 7 on land use. But it is important to emphasize again that water quality and most land practices are tied closely together. Wherever silt is a problem, water quality provisions must extend to land use. Filling in a brook or replacing it with a culvert must be stopped if the result is going to be water pollution from sediment.

WATER QUALITY
STANDARDS

The 1965 and 1970 Water Quality Acts provided the basic clean water standards. The 1965 Act established the Federal Water Pollution Control Administration, whose name was changed in 1970 to the Federal Water Quality Administration (there is wishful thinking that one of these days pollution will be prevented, not merely controlled). Previously a 1956 bill had given the federal government vague means to set standards for interstate and coastal waters. A 1966 bill expanded significantly the program of federal grants to municipalities that raised at least 30 percent of the cost of waste treatment. Now, federal aid of one form or another is available for construction of facilities through the standard FWQA program: Housing and Urban Development, local government grants for hard-pressed communities of under ten thousand population, the Economic Development Administration grants to "depressed areas," and Farmers Home Administration rural aid to small communities that are not part of a town with more than 5,500 population. Some states have developed assistance programs as extensive or more so than federal aid. Examples are New York, Maryland, Massachusetts, Michigan, Ohio, Pennsylvania, Oregon, Washington, and Wisconsin. Other states have clean water programs but it is notable that their governors have not requested funds to implement these antipollution activities even though the state programs anticipate federal matching grants.

Under the existing legislation, the states are to set water quality standards, and propose plans to implement them in accordance with federal criteria. The best reference for these

guidelines is the 1968 report entitled "Water Quality Criteria."
It set standards of cleanliness, depending on what classification
a state gave to a waterway. For example, if the water was a
drinking source or was desirable for fishing, criteria for bacteria,
temperature, dissolved oxygen, suspended solids, and so forth
would be far more restrictive than if the water was classified
as an "industrial river," which is a euphemism for an "open
sewer." In theory, if a state does not enforce its federally approved standards, the Federal Government can intervene. In
fact, and in accordance with policy encouraging state responsibility, federal authorities have only initiated enforcement actions
in interstate waters or when pollution effected a product marketed in interstate commerce. Otherwise, the FWQA awaits
invitation by a state governor to assist in a cleanup effort. The
legislation should be studied carefully, for as in all laws there
are myriad possibilities for the astute, persistent activists. Copies
of the bill can be obtained from the various agencies, from the
House and Senate Public Works Committees, from state water
resource boards, and congressional representatives.

Presently there are proposals before Congress that would
obtain even stricter standards, tougher enforcement, more incentives for construction of water treatment and recycling systems, and even greater public participation. It is generally
agreed that the federal authority should apply even to waterways within state boundaries and serious thought is being given
to the regulation of water quality through emission standards.
Then rather than judge a city or industrial polluter according
to the degree his discharges degrade water quality, the effluent
itself would have to meet criteria and would in effect be compared to acceptable water.

There is not space in this book to explore the subtleties and
possibilities for manipulation in existing water quality legislation. Nor would it be productive here to argue the costs of
cleaning up because estimates vary so tremendously. The Nixon

administration's figures *do* seem to be conservative compared to those of Maine Senator Edmund Muskie, who has gained expertise as chairman of the Senate public works subcommittees on air and water pollution and solid waste disposal. President Nixon proposed a federal contribution of four billion dollars to a total commitment of ten billion dollars for waste treatment facilities over the next five years. But because of a budget device, the administration's four billion dollars could be stretched out over eight years. Senator Muskie's water package is two and a half times as big. He has proposed federal grants totaling 12.5 billion dollars over five years, matched by state and local government.

All kinds of figures are bandied about when the costs of cleaning up are discussed and, as already noted in this chapter, no one is certain what strategy the battle will adopt. It is estimated, for example, that it would cost fifty billion dollars to separate storm sewers from main lines. However, it is not determined that separation is the answer to city systems that bubble together when overloaded. Possibly it will be more efficient to provide combined treatment, provide storage systems to hold rainfall, or build lakes to be filled by treated storm sewage and used for recreation, as is currently planned by Interior Secretary Walter Hickel in the Anacostia River. The recreational benefit would make treatment of the storm sewage worthwhile.

Then it is estimated that the nationwide installation of tertiary treatment plants would cost per capita about 10 percent of the average citizen's investment in getting basic utility services—electricity, telephone, and water—which add up to about $300. But as already noted, tertiary treatment may be a waste of money when water of secondary quality can be reused.

Taking another tack, Wisconsin Senator William Proxmire has proposed an "effluent charge" to prod industry into installing pollution controls at its own expense. His bill would set charges on effluent according to a factory's volume of discharges into a waterway and according to the severity of the pollutants. In other words, the factory would be rationed according to what its water body could absorb. The proposal, made flexible to allow for reshaping in the legislative process, would levy effluent charges higher than the cost of investing in controls. Moreover, revenues from the charges would, according to Proxmire, provide the federal government with a steady annual flow of up to two billion dollars, half of which would be granted back to cities for waste treatment construction and half of which would go to regional organizations, such as river basin commissions, for water quality programs. One point always can be made in weighing the costs of clean water. Whenever estimates are broken down per capita, including both the cost of clean-up and future prevention, the price for water quality is cheap compared to what is spent on cigarettes, liquor, and (giving bad habits a break) far less essential indulgences.

In the city of Washington, it would cost two cents per person per day to install full secondary waste treatment, another one cent to have phosphorus removed, and two more cents for tertiary treatment—waste water made drinkable again.

In sum, if you are in the throes of wondering what aid is available to combat water pollution, first study existing legislation. The League of Women Voters, which successfully led the 1969 Citizen's Crusade for Clean Water, has published first-rate material on water quality programs and alternatives. (See Bibliography.) The League has also established a successful record in citizen action for sound water resource management throughout the nation by instigating river basin clean-up campaigns, pushing bond issues, or promoting local initiatives for antipollution outlays and action.

It should be quite apparent by now that water is a resource whose quality depends on an extraordinary number and variety

of conditions. Pollution control is only one aspect of water
quality. The way water is moved or stored and the use to which
it is put must be thoroughly studied before pollution per se is
given attention. It doesn't make sense, for example, that the
city of New York not recycle the Hudson or experiment with
desalting techniques to obtain drinking water. The city now
relies on an awkward and old-fashioned network of dams,
reservoirs, and aqueducts for its supply. And to keep this system functioning the city must continuously, and most delicately,
barter with communities whose water it is pirating. Needing
1.3 billion gallons a day, the city's consumption has so lowered
the Neversink River that it belies its name and looks like a
great channel of slime much of the year. Of the 150 gallons
a day used by each New Yorker, only thirty are essential for
cooking, cleaning, and hygiene. Water rationing ought to be
tried, or New York should set its rates high enough to insure
water economy. (New York Mayor John Lindsay indeed requested and obtained water rates 75 percent above previous
charges, but cutting down water consumption was not given as
a reason and it remains to be seen whether the new rates induce
water savings.)

It has been suggested in one form or another that water
resource agencies be given the power and scope to rule on every
aspect of water management rather than cope with just bits
and pieces of the overall problem. At the federal level, the
FWQA should have the power to correct land abuses that lead
to water pollution, should plan ahead for the nation's water
needs, and should manage flood control, navigation, and
stream "improvement" projects. It should be one agency, not
several with conflicting interests, that continuously reviews all
the activities that influence water quality and that can move fast
to correct abuses. Reorganization of the FWQA and other environment agencies is presently underway. However, it is not
yet clear what the scope of the new Environmental Protection

Agency will be with respect to the full range of water management activities.

On a regional basis, watershed and river basin management is a promising development. However, thus far, none of the interstate river basin bodies has much leverage. None but the Delaware River Basin Commission has more than the authority to *plan* the development, use, and protection of a river system. The Delaware Commission, encompassing streams and estuaries in four states, has actually set water emission standards and quotas that are enforceable. But the real test comes when the Commission has a chance to move forcefully to transcend the vested interests of its various water-consuming constituents. That test has not come. Even so, river basin commissions have the potential to demonstrate whether regional government can be successful. And if they work, it could be that the apparatus of environmental action will decentralize and take on a personal and participatory character throughout the nation.*

ACTION LIST

There is no end of suggestions for things to do these days to obtain an ample and clean water supply. You can put bricks in the water closet to reduce the volume of water (seven gallons) that flushes the toilet. You can refuse to buy any number of products that end up as pollutants—from phosphate detergents to plastic or aluminum containers (see Chapter 5) which forever resist decay and absorption by the aquatic environment.

But first it would be more productive to learn the charac-

* Since this was written, the President's Environmental Quality Council has recommended that a river basin be designated a demonstration project to explore "the most advanced concepts of water quality management."

teristics of your water environment. Ask these questions of government officials (and when it appears that an answer may be politically expedient or weighted to favor an economically important polluter, then query local conservation organizations or university biologists for additional facts). How much water is used domestically, by manufacturing and by electric utilities? Where does it come from? What standards are required by various users? How efficiently is it used? What is the present state of the drinking water? How is it treated? Where does it come from? What are the possibilities for recycling waste water or using it for fertilizer? How are gasoline, oil, and other hazardous materials being disposed of? What are the community's laws and physical arrangements for managing water runoff and preventing silt erosion? In general, how does land use affect the watershed, and what zoning laws exist to maintain a sound land-water relationship? Where does the state stand in enacting water quality standards to meet federal criteria? How do they compare with federal criteria (see sources in Bibliography) for temperature, flow, dissolved oxygen, hydrogen, salinity, silt, light and turbidity, nutrients, toxic substances, pesticides, slimes, and nuisance organisms? Are there hearings scheduled regarding these standards? Finally, what are the possibilities for river basin programs in your region?

Water has a generous capacity for recovery from abuse, particularly if it is flowing. But the cures are not easy or cheap, nor will they come about without individual questioning and action. The slightest commitment helps.

2 ⚜ AIR

IF WATER IS VISIBLY our most degraded natural resource, the most ominous and pervasive threat to environmental quality and human health is air pollution. The ambient, or all-encompassing air that we breathe respects no boundaries. It rides the winds or becomes overburdened with noxious gases when not dispersed by weather changes, so we have to take it as it comes. Once the air is polluted, unlike dirty water, it cannot be purified unless of course it is to be contained in a sealed space. Some of the harmful contents are scrubbed out by rainfall, but much remains suspended above us, producing long-term effects that we wish we knew more about. What *is known* is terribly upsetting. For example, a nonsmoker walking the streets of New York—to and from work and during meal hours—probably will breathe into his system carcinogenic pollutants more than equal to a pack of cigarettes. And he really has no choice. When air pollution is perceptible, it must be assumed that also present in the air are all kinds of invisible poisonous gases—such as carbon monoxide—that make up 90 percent of the health-endangering pollutants.

For centuries, the lethal effects of air pollution have produced the most spectacular tragedies. Public concern over soot-blackened skies led in 1273 to the first antipollution ordinance in London, England. But the English air only got worse as the island's population grew and the industrial revolution's furnaces added more smoke. Bronchitis became known as "the national disease." In more recent times (throughout four days in 1952), a temperature inversion—a layer of warm air sitting above and trapping cool air on the ground below—produced a poisonous fog held responsible for pushing the number of deaths four thousand beyond the usual rate for that period.

Not until 1956 did the English finally pass a Clean Air Act that was effective *and enforced*. It restricted home burning of coal, proscribed smokeless fuels, recommended general precautionary measures, and set up a system of alerts to avoid a repeat of the 1952 disaster. The results have been dramatic. The city of London issued a report in 1970 showing that, at a *cost per citizen of only thirty-six cents a year,* respiratory attacks had dropped, sunshine during the winter home-heating period had increased 50 percent, and the degree of visibility had tripled due to smoke reduction. Great public buildings and monuments were not nearly as corroded and blackened, and during this period the number of bird species more than doubled. In sum, while London has been losing the romantic fog that Ella Fitzgerald used to sing about, it has regained its health.

In the United States, public outcry began to mount in California when Los Angeles in the early forties was perpetually enshrouded in smog. But concern was not widespread until in 1948 an inversion over the steel and chemical town of Donora, Pa., sickened over half the fourteen thousand inhabitants and claimed twenty-three lives. Then in New York, over Thanksgiving of 1966, an estimated 168 "excess" deaths were blamed on a choking smog. As a result of these, other incidents, and a dedicated public relations campaign by U.S. health officials, the public in recent opinion polls has labeled air pollution the number one threat to the quality of the environment and human health. But still, not a day passes when citizens somewhere are not coughing and complaining. The condition that the moguls of the automobile industry once said was uniquely Los Angelean has beset hundreds of cities that have similar climactic phenomena and has even invaded pastoral valleys in the nonindustrial northern half of New England and other rural regions. A pall of soot, smoke, or smog is hanging over some city every day. In the country plains or pastures, the heavy haze may have wafted from cities miles away or, just as likely, from a local factory (in which case, little is being done about it because the source is apt to be the region's only or major employer).

I have done my share of coughing, sniffling and eye-rubbing in the air-sick cities, but nothing was quite so impressive as seeing the breakup of a Phoenix smog one morning in late March, 1970. For several days an inversion had allowed discharges from Arizona's copper smelters to build up and then converge over the city and environs. Added to that was smog caused by the reaction of automobile exhaust to sunlight. Yes, the region's main attraction, sunshine, had become an accomplice to air pollution. Then the inversion dissipated as a cold front was ushered in by a southeasterly wind. That was fine for Phoenix, but sixty miles to the northwest, where at dawn the air had been crisp and clear, the effect was eerie. Down the desert plain, between mountain ranges that had stood out in sharp outline moments before, came a thick brownish billow of dust that made the jackrabbits and quail hyperactive. But this was no dust storm, because the wind had blown harder from the opposite direction the day before. It was just the man-made pollution from around Phoenix that was passing through.

One does not have to be so visibly struck to become aware of the debilitating effects of air pollution. Clinical evidence, epidemiological studies, and data collected by monitoring stations are frightening. Ecologist Barry Commoner does not exaggerate when he says so often that the present generation is the first to grow up with strontium 90 in their bones, DDT in their fat, and asbestos in their lungs. A year from now, he might have to expand that statement. One fact stands out. Grave danger to human health is presented by the ambient air constantly in some sections of the country and at varying frequences everywhere else. C. C. Johnson, head of the Environmental Health Services of H.E.W., said in a speech in

May, 1970, that "The current health costs of air pollution
alone are estimated conservatively at $4 billion per year. . . .
Chronic respiratory disease, which in large measure can be
associated with occupational conditions and air pollution, is
the nation's second leading cause of disability. Social security
payments to victims of this disease and their families total $90
million annually. Emphysema, the major chronic respiratory
disease, causes nearly 50,000 deaths per year."

What is difficult to obtain and what too often eludes scientists
and doctors is proof that one particular pollutant or another is
the cause of sickness or death. And when the evidence is
gathered, tracing it to a source is often just as trying. Like the
agents that interact underwater, the components of the air have
synergistic effects upon one another or they react to new elements such as rain. Sulfur dioxide, for example, combines with
water droplets or fog to form sulphuric acid mist that damages
human lungs and literally corrodes buildings and monuments.
Because air pollution is so hard to qualify and measure, combating it has been, and continues to be, a sad and often pathetic
tale. As we shall see, present approaches and trends will have
to be reversed rather abruptly if the story is to have a happy
ending. And because citizen indignation—and participation—
will be essential to turn the tide, you should have a basic
knowledge of types of air pollution and what they do to you.

AIR POLLUTANTS

The main air pollutants—and the list grows as more are
understood or are found to be more prevalent—are sulfur
dioxide, hydrocarbons, carbon monoxide, nitrogen oxides, and
particulates. A 1968 estimate revealed that 214 million tons of
these substances had soiled the ambient air over the U.S. Other

pollutants such as ozone, asbestos, lead, berylium, radioactive isotopes, hydrogen fluoride, various oxidants, and a tremendous variety of chemicals such as herbacides and pesticides, are just as harmful. As the so-called "environmental decade" of the seventies got underway, only a few cities and counties (e.g. New York, Chicago, and San Francisco) had laws for policing their skies, and federal air quality criteria were not yet implemented or enforced.

Sulfur Dioxide

This is the pollutant associated with the big disasters in London, Donora, New York, in the Meuse Valley, Belgium (where hundreds were sick and sixty died in 1930), and in other congested urban areas. Sulfur dioxide is a product of the combustion of sulfur-bearing fossil fuels such as coal and oil. Electric utilities contributed nearly half of the 33 million tons in 1968. The other important contributors were large industries, heating furnaces, and cars and garbage incinerators.

Sulfur dioxide is quite likely the cause of considerable ill health. Available evidence indicates that people exposed to it from birth suffer more than those who are not. It is *known* to aggravate respiratory diseases that chronically afflict older people, such as emphysema and bronchitis, by damaging the bronchial tubes of the upper respiratory system or infiltrating the air sacs (alveoli) in the lungs of the lower respiratory system, where the blood receives its vital supply of oxygen. In either case, breathing becomes more difficult and an added load is placed on the heart. In England, bronchitis and emphysema are the major causes of death. Emphysema is the fastest growing killer in the U.S. On the heels of English studies showing a correlation between sulfur dioxide emissions and sickness or death, U.S. investigators are finding increasing instances of sulfur dioxide poisoning. One study showed that sulfuric acid mist increased "airway" or breathing resistance as much as four times. While the sulfur dioxide criterion or acceptable level set by the Department of Health, Education and Welfare is an annual average of .04 parts per million, it is known that this amount is exceeded all the time in most cities and runs several times that level in certain areas. HEW studies have shown that human health suffers when the average level of sulfur dioxide is higher than .04 ppm and vegetation is affected at .03 ppm. A monitoring survey in 1968 revealed that Chicago had an "an annual mean concentration" of .116 ppm of sulfur dioxide and Philadelphia .08 ppm. New York's average is currently .06 ppm.

Aside from what it does to your health, sulfur dioxide is a major destroyer of property, particularly in the form of corrosive sulfuric acid mist. In England, the Beaver Committee in the early fifties estimated that air pollution property damage ran well over $700 million, or upwards of five pounds ($14) per person. In some areas, 125 tons of pollutants fell per square mile, killing crops and plants. The process of photosynthesis by which greenery uses solar energy to convert carbon dioxide back into oxygen was inhibited in the thickened atmosphere, and pollutants clogged plant pores to prevent transpiration of water—an essential phase of the photosynthetic cycle. According to U.S. officials, sulfur dioxide from a Virginia Electric Power Co. plant on Mount Storm, West Virginia, has virtually wiped out commercial Christmas tree plots in nearby Maryland. While much of this sort of damage can be blamed on particulate matter, as will later be noted, it is often immaterial to differentiate between sulfur emissions and the soot that is a kind of particulate. The Citizens for Clean Air, Inc., have shown New York's property damage figures in a most effective advertisement. It is an aerial photograph of New York on which the annual dollars of damage per family are inscribed for various districts, depending on the amount of fallout each area receives. In Manhattan, a family of four would have to spend up to $850 to clean up and repair the damage done by dirty air. Westchester and Staten Island families get off easier at $580.

Property damage from sulfur dioxide and particulates runs upwards of $12 billion a year in the U.S., breaking down to more than $65 per person. Perhaps the most famous example of what the ambient air can do is the erosion of Cleopatra's Needle in Central Park. The windward sides of this obelisk have been eaten away several inches. The hieroglyphics that were unscathed when Egypt gave the 1600 B.C. monument to the U.S. ninety years ago have thus been all but obliterated.

Hydrocarbons

The gaseous product of incomplete combustion, hydrocarbons are the emissions that along with oxides of nitrogen, and in concert with sunshine, produced the first smog over Los Angeles. At that time there were only thirty million automobiles in the entire U.S., far less than the nearly ninety million that have helped spread the malaise to other cities today. Cars produce more than half of the 32 million annual tons of hydrocarbons. Another heavy contributor of hydrocarbon tonnage is the petro-chemical industry.

The adverse health effects of hydrocarbons are not well diagnosed except in one instance. The most potent hydrocarbon, benzo-a-pyrene, is present in cigarette smoke and is associated with cancer. It was described in a 1968 article in *The New Yorker* magazine by Edith Iglauer, which was considered so helpful in shedding light on all the biological effects of air pollution that it has been reprinted for distribution by HEW (see Bibliography). She cited one study suggesting that benzo-a-pyrene acted synergistically with sulfur dioxide. Certainly both pollutants are present in strength in urban areas where lung cancer rates are high. As previously noted, the man on the street is assailed by more benzo-a-pyrene than he inhales from a pack of cigarettes. The "Task Force Report on Air Pollution" by Ralph Nader's Center for the Study of Responsive Law (see Bibliography) cited data compiled in Los Angeles showing that benzo-a-pyrene was directly carcinogenic in test animals. And Senator Edmund Muskie's subcommittee on air pollution was presented with evidence strongly suggesting that benzo-a-pyrene from cars was responsible for unusual increases in lung cancer deaths among animals and birds at the Philadelphia Zoo, which is compressed between two heavily traveled thoroughfares. "A Report of the Senate Republican Policy Committee," the first of a series of 1970 papers, matter of factly puts down hydrocarbons as a likely contributor to increased lung cancer among urban populations.

As for hydrocarbons contributing to smog, the process and consequences are quite thoroughly diagnosed. Not only does smog irritate one's eyes and throat but it reduces visibility seriously enough to have become a safety hazard around every large airport. In addition, by cutting off sunlight, smog has become one of the nation's major crop destroyers. Exact estimates of damage vary and, unless they are for a specific spot it is hard to ascribe the damage to just one instead of several pollutants acting together. Anyway, national agricultural losses due to air pollution are estimated to run nationwide over $500 million. Such damage, mainly from smog, runs over $130 million a year in California. Leafy vegetables such as lettuce and spinach are just not grown around Los Angeles anymore.

Oxides of Nitrogen

This pollutant is the other major automobile emission that helps produce photochemical smog. Cars contribute just less than half of the 20 million annual tons. Electric utilities, metal fabricating, and chemical plants are the other big sources. Oxides of nitrogen naturally become nitrogen dioxide which in turn, after absorbing the sun's ultraviolet rays, becomes nitric oxide and atomic oxygen. Interacting with hydrocarbons, these oxides of nitrogen produce photochemical smog and a by-product poison, ozone. Above levels of .3 ppm, ozone restricts breathing and hurts the eyes and throat. It has already been diagnosed as the cause of affliction of 1.3 million acres of stately ponderosa pines in the San Bernadino Valley eighty miles east of Los Angeles. Some of these trees have died and the rest are considered so far gone that the U.S. Forest Service has been leasing the pine forests to commercial loggers. The University of Massachusetts pumped ozone into an experimental green-house at the same level that is found in Boston traffic on a sunny day. Carnations and geraniums, the test flowers, withered away and produced from a quarter to a third less new growth. Another synergistic union of oxides of nitrogen and hydro-carbons produces peroxyacl nitrate (PAN), a component of photochemical smog that irritates eyes and damages mucous membranes.

U.S. air quality officials are finding increasing reason to as-sociate oxides of nitrogen and respiratory ailments and they note that accidental high-level exposure has caused fatal pneu-monia. City children exposed over long periods to low levels of oxides of nitrogen have been found more susceptible to colds than their country counterparts. A disturbing feature about this pollutant is that, as the auto's internal combustion engine is improved to reduce hydrocarbon and carbon monoxide emis-sions, oxides of nitrogen increase substantially in the exhaust. This phenomenon is one of the main reasons scientists, air pollution fighters, and citizen advocates in Congress insist that the present car must be replaced by a totally different propulsion unit instead of being merely bandaged in its conventional form.

Carbon Monoxide

Still another product of the car (which contributes over 60 million of the annual 100 million tons), carbon monoxide is a pollutant that is quite well understood as it affects human health. Late in 1969, the National Academy of Sciences issued findings on carbon monoxide, one of which was that sustained exposure to levels of 10 ppm or above produced measurable adverse effects upon human beings. This level is well exceeded during peak commuter hours in every city and throughout the day in congested metropolises such as New York, Chicago, and Los Angeles. Carbon monoxide passes without warning, unnoticed, into the blood system. It combines with hemoglobin to impede the transportation of oxygen from the lung to the body tissues. Such deprivation of oxygen slows down reactions, produces nausea and dizziness, blurs visual acuity, and affects the brain generally. This knowledge should be no surprise. After all, for years people have been committing suicide by sealing themselves in the garage and running the car, or by getting inside the car after taking a hose from the exhaust pipe through the window. In an informative workbook (see Bibliography), the Scientists' Institute for Public Information of New York inserted a table showing sample carbon levels in cities. Along the freeways of Los Angeles, the level averaged over 30 ppm's and as high as 54 ppm; in a parking garage, 59 ppm; at a Cincinnati intersection, 20 ppm; for a brief interval in Detroit, 100 ppm, and at the average Manhattan intersection, 15 ppm all day long. Even New York's air quality goal of 15 ppm would allow visual handicaps, slowed reactions, and some loss of judgment, if tolerated for long, the scientists said. The National Academy report cited many studies correlating carbon monoxide exposure to increased sickness and death rates among respiratory sufferers and people with heart conditions. The study also noted that women in pregnancy are particularly vulnerable to oxygen deprivation because of their additional burden. It is abundantly clear, though difficult to pinpoint, that many of the traffic accidents along congested highways are due to the crippling of driver responses by carbon monoxide ingestion.

Particulates

Such is the classification of the particles of soot, fly ash, the fallout from all kinds of industrial activity and the unburned specks that pour out of smokestacks and incinerators and appear every time a fire is lit in a dump, field, woods, or backyard. For some time it was thought that particulate matter was unable to penetrate the protective apparatus of the human breathing system and was a problem only as it blackened buildings, soiled the wash on the line, begrimed windows, and coated one's clothing. But now available evidence indicates that small particulates carry sulfur dioxide, DDT, and other harmful agents into your respiratory passages. New York's particulate dustfall is tremendous, seventy thousand tons in 1969 and in some spots up to thirty tons per square mile per month.

Particulates have played protagonist in a debate that continues to perplex the ecologists. Carbon dioxide is the other key player and the question is which one will win in changing the temperature of the biosphere to bring on a new ice age or a great flood. It goes like this. Worldwide combustion since the industrial revolution is believed to be responsible for a slight increase in global temperature up until 1940. The reason is that carbon dioxide, another product of combustion (as well as breathing), acts as a heat radiative by retaining some solar energy that would otherwise return to outer space. If the earth were to warm up only a degree or more, the polar ice cap would begin to melt, raising the level of the oceans enough to flood most of the world's great cities. However, a slight decrease in the earth's temperature since 1940 is laid to the buildup of particulates that hang over human settlements like fogbanks. Not only does this blanket keep out some sunlight but dust particles help to form clouds and start rains by attracting water vapors. Meteorologists say this is the reason the cities receive more rainfall than their environs. Evidence regarding this theory was revealed by scientists of the Smithsonian Institution. Their surveys showed that the amount of sunlight reaching the ground in Washington had diminished 16 percent compared to readings taken fifty years before in 1909. Since the measurements were taken on clear sunny days, air pollution, not weather trends, was held responsible. If the air pollution trend were to continue, some say the resulting temperature decrease could cause another ice age as soon as the next century and the large cities would be scoured clean by glaciers. Yet still another scenario unfolds in the wings—the increasing destruction of open spaces and green belts where plant photosynthesis converts carbon dioxide back into oxygen. If that rate continues, coupled with growing fuel combustion, ecologists predict the volume of carbon dioxide in our atmosphere would increase dramatically and the warming trend would resume. Either way, *we lose,* unless you consider the result the most effective way of solving the population crisis.

The Nader team, directed by lawyer John Esposito, cited studies suggesting that particulate concentrations in cities contributed to cancer of the stomach, bladder, esophagus, and prostate. In coal and steel producing Pennsylvania, citizens in 1969 were most unhappy with the state's proposal to set particulate criteria of 100 micrograms per cubic meter with a goal of 80 micrograms, the HEW recommended level. A coalition, dubbed the "Breathers' Lobby" by the *Wall Street Journal,* fought successfully to get a criteria level of 65 micrograms.

In their stand, they cited increased death rates among people over fifty, blamed on exposure to both particulates and sulfur dioxide.

Lead, Asbestos, Berylium, and Others

Like carbon monoxide, lead in the air is known, at a certain level, to kill. In older, poorer sections of cities, children die or suffer brain damage quite regularly from inhaling lead particles that flake off walls that still bear lead-based paints (now outlawed). Doctors are disturbed to find that the human body retains the same percentage of lead traces as are found in the immediate environment in which that human lives or works. It is thus evident that as the atmospheric lead level rises there is real cause for alarm. Today the major contributor of lead is, you guessed it, the car. The Nader report noted that the annual contribution of the auto was 400 million pounds. The lead particles spew out of tailpipes as unburned gas additives considered essential in the fuel to prevent "engine knock." The auto industry claims that the current crop of engines will not run efficiently on nonleaded gas, although many motorists, including myself, are happy with the only nonleaded gas on the market that is produced by AMOCO, actually as aviation fuel, in a refinery purchased as surplus from the government. This year the federal government set air quality standards that would force the auto and oil industries to collaborate on a clean fuel without lead additives. The auto makers say they are not unhappy because lead wears out antipollution exhaust devices and will ruin the catalytic mufflers that will be necessary to cope with new oxides of nitrogen standards. The fuel makers say their changeover will be burdensome, although the government has estimated that the excess cost of lead-free gas passed along to the driver would be about one cent per gallon initially, and maybe less as the change works out and if the internal combustion automobile survives. Of course, there would also be a *saving* on engine and sparkplug wear.

Another known killer is asbestos. The major source—besides the factories in which it is made and where it is an occupational health hazard—is the construction industry. The air on the streets of every modern city is full of asbestos dust, fallout from the fireproof spray applied throughout rising buildings. New York has put a temporary stop to asbestos spraying, ordering new ways of putting it on or the use of fiberglass substitutes. Another significant source, again, is the car. Asbestos is used in brake linings and clutch facings, and particles of it are dispersed as brakes and clutches are worn down.

What does asbestos do to you? Its very indestructibility makes it persist for a long time after it enters your lungs. Cancerous growths are apt to form as the lungs try to get rid of the asbestos. U.S. health data show that one out of every five asbestos insulators dies of lung cancer, seven times the national average.

The list of air pollutants is endless. As disturbing results pop up, yet another culprit is named. It would not be productive in this chapter to run down more than a few—and their effects—since the moral of the story is so simple and basic. It was applied previously to water. Presume that a combustive product or dust-producing substance pollutes until you know otherwise or have built controls into the combustion system. Dr. Raymond Slavin of St. Louis put it quite well during air pollution field hearings. "No one should release any substance into the air other people must breathe unless it can be shown to be harmless, unless he can prove that it is not injurious," he said, and added, "This is the guiding principle in the acceptance of food additives and drugs."

But just for the record, here are some of the "other"

pollutants that have been allowed to escape: toxic hydrogen fluoride exhaust from rockets; fluorides from phosphate fertilizer plants that have caused thousands of cattle to die in Florida by destroying their vegetation; berylium used in making metal alloys that causes a debilitating form of lung infection; persistent pesticides that upset all living organisms and may be carcinogenic to humans; herbacides like 2,4,5-T that are supposed to kill only brush but produce birth defects in animals and humans as well; radioactive particles that may emanate from any nuclear project and cause thyroid damage and cancer, and ethyleneimines from insecticides and industrial processes that have deformed test animals. When industries use chemicals in their manufacturing processes and when chemicals themselves are being formulated, without proper precaution there is certain to be some noxious by-product blown into the ambient air and the chances are that you won't know it until there are "excess" sicknesses or deaths.

A Partial Guide to Industrial Air Pollution

The following table compiled by the American Chemical Society, provides a guide to the usual emissions associated with specific industries. The order of the list does not necessarily indicate the rank of amounts or the seriousness of the emissions. If you live in the vicinity of any of these kinds of industry, it might be worth checking with the companies in question to find out what they have done to reduce these and any other emissions.

SOURCE	TYPE EMISSION
Integrated steelmills	Particles, smoke, carbon monoxide, fluorides

Nonferrous smelters	Sulphur oxides, particles, various metals
Petroleum refineries	Sulphur compounds, hydro-carbons, smoke, particles, odors
Portland cement plants	Particles, sulphur compounds
Sulphuric acid plants	Sulphur dioxide, sulphuric acid mist, sulphur trioxide
Grey iron and steel foundries	Particles, smoke, odors
Ferro-alloy plants	Particles
Kraft pulp mills	Sulphur compounds, particles, odors
Hydrochloric acid plants	Hydrochloric acid mist and gas
Nitric acid plants	Nitrogen oxides
Bulk storage of gasoline	Hydrocarbons
Soap and detergent plants	Particles, odors
Caustic and chlorine plants	Chlorine
Calcium carbide manufacturing	Particles
Phosphate fertilizer plants	Fluorides, particles, ammonia
Lime plants	Particles
Aluminum ore reduction plants	Fluorides, particles
Phosphoric acid plants	Acid mist, fluorides
Coal cleaning plants	Particles

AIR POLLUTION LAWS

Two observations must qualify a discussion of air pollution laws before it even begins. First, while some states have passed antipollution ordinances, the laws are not tough or are not enforced. California has taken the lead in controlling cars' emissions but the last battle is far from being fought, particularly as highway planners go on encouraging use of cars instead of transportation alternatives. Pennsylvania has proposed tough air quality criteria but the battle will come when enforcement procedures are worked out and followed up. Second, while the federal government has been empowered to deal with the problem since the 1955 Air Pollution Act authorized five million dollars annually for research (through 1960), there is not yet in existence *a really enforceable* air quality plan under U.S. law except for automobiles, and this has given Detroit far more latitude than it deserves. There have been additional federal air quality acts or amendments in 1960, 1962, 1963, 1964, 1965, and 1967. The last two are those that presently guide efforts to clean up the air.

The 1965 legislation consisted of amendments to the 1963 Clean Air Act. One amendment was the Motor Vehicle Air Pollution Control Act, allowing HEW to set emission standards for cars. The other was the Solid Waste Disposal Act, providing research and development funds to find clean alternatives for getting rid of all the refuse that is now burned or used as land fill (thus often polluting water).

The Automobile

It is easy to work oneself into a lather over the attitude of the automobile industry toward air pollution—not to mention highway safety. Time and time again, Detroit's engineers and their bosses have claimed the car was an inconsequential polluter or that the problem was all but solved, just as independent or government scientists kept right on compiling evidence that such contentions were pure bunk. The record shows that Detroit has not yet ceased to indulge in this sort of optimism, seemingly conceived as a holding action to keep us all at bay still another year. Since the industry has been obstructive toward alternatives, we must continue to drive their polluting cars and so the delaying tactic works.

Yet by now it is well publicized that the car is responsible for 60 percent of the annual tonnage of air pollutants. Not only is it a health hazard but it hogs land space and congests urban arteries, bringing untold grief to those who live near highways and those who must travel them. All the same, in April, 1970, Charles Heinen, the incurably optimistic Chrysler chief engineer for emission control research, said that by meeting 1971 standards the car would no longer be a significant polluter. After that, he was saying in speeches, "Why waste money on further controls on autos and jeopardize other programs that are more important?"

The following table shows automobile emission standards now and those set for future years:

MOTOR VEHICLE EMISSIONS
(grams per vehicle mile)

	1970	1971	1973	1975	1980 *
Hydrocarbons	2.2	2.7	2.7	0.60	0.25
Carbon monoxide	23.0	23.0	23.0	11.50	4.70
Nitrogen oxides	None	None	3.0	0.95	0.40
Particulates	None	None	None	0.10	0.03

* Tentative.

It is questionable whether the auto industry can achieve these goals with the conventional internal combustion engine. Now under consideration are recirculating devices for oxides of nitrogen, lowering the combustion temperature in the cylinder to reduce oxides of nitrogen, highly refined carburetors or fuel injection systems, thermal reactors made of heat-resistant metal instead of manifolds, and new fuels without lead. But even if the 1980 emission standards are met, the total tonnage of pollutants is expected to rise again because of the predicted population increase in cars.

There are several alternatives to the internal combustion engine. And it is important to note that, despite industry disclaimers, very thorough independent studies invariably conclude that some of the proposed replacements for the internal combustion car are quite feasible and that all require much more attention than either government or industry has spared up to now. Federal studies estimate that it would cost $50 million to develop a sound prototype alternative. This is about one-fifth of General Motor's advertising budget and a pittance compared to the annual estimated one billion dollars spent by the auto industry in retooling for style changes. To be sure, a prototype is only the beginning but most outside experts feel that by postponing the annual styling ritual, which is unnecessary whimsy, industry could well afford to develop production line clean-air novelties.

A 1969 staff report for the Senate Commerce Committee titled "The Search for a Low Emission Vehicle" (see Bibliography) concluded flatly, about one much touted novelty, that "The Rankine cycle propulsion system is a satisfactory alternative to the present internal combustion engine in terms of performance and a far superior engine in terms of emissions." The Rankine engine works on the steam principle. Pistons or turbines provide the power and are driven by steam vapor heated in a coil. While low-grade gas or kerosene provide the heat, exhaust emissions are fractional compared to the present car. The internal combustion engine is, after all, a most inefficient fuel consumer because combustion consists, in effect, of a continuous series of explosions. In the Rankine system, *external* combustion comes from a steady flame which, in addition to burning more cleanly, is consumed more slowly.

Gas turbine engines are already available for trucks and buses and they are far cleaner, quieter, and even more efficient than diesel units. But there are innumerable obstacles to scaling down the gas turbine to a car-sized unit. Other possibilities which appear further off than the Rankine engine are hybrids (e.g., using electricity for short, city distances and a combustion unit for the freeway) and electric cars powered by a fuel cell.

Conventional engines can also be converted to burn propane or natural gas. For a fleet of cars (e.g., taxis or municipal vehicles) the cost runs around $350 a car. However, the nation's supply of natural gas presently is considered too limited to make this alternative feasible on a larger scale. Moreover, natural gas is provided in vapor under pressure and thus the amount that can be carried limits cruising range. Liquid natural gas poses a safety hazard because of its high volatility. However, propane conversion is quite suitable for city driving, particularly by a number of cars under one management so that costs are minimized and the fueling arrangement centralized. If they are driving outside the city or if they run out of propane they can switch back to conventional fuel.

Nothing short of public indignation can do anything about the car. It is painfully apparent that the auto and fuel establishment and their satellites, the service and repair industries, as well as the highway builders and their dependents, the cement and asphalt producers, still hold the upper hand in perpetuating an inefficient and troublesome creature that will hold us in its gaseous spell for a long time to come. The Nader report contains a documented and devastating look at Detroit's foot-dragging and is recommended reading on that score alone. Interestingly enough, while the report is scathing in its review of HEW efforts, many of those who bear the brunt of criticism privately agree that the auto industry has been getting off easy at the hands of the public and government.

One final note supports that point. The figures for auto-

mobile pollution are based on a number of premises, one of
which is that the present pollution controls function according
to specifications. That is wishful thinking. Just to see, HEW
ran tests on rental agency car fleets in Los Angeles and Detroit. The rate of failure varied by make of car but went higher
than 70 percent.

Here is the breakdown:

RENTAL CAR EMISSION TESTS
(percentage of vehicles failing to meet standards)

MODEL	HYDRO- CARBONS	CARBON MONOXIDE	BOTH	EITHER
American Motors 290 cu. in.	0	8	0	8
Chrysler, 318 cu. in., 2bbl carburetor, 9.2 compression ratio	13	10	4	19
Ford, 302 cu. in., 2bbl carburetor, 9.5 compression ratio	57	18	17	58
Chevrolet, 307 cu. in.	73	44	44	73

The California Air Resources Board conducted a similar survey and found that failures increased dramatically as cars put on
more mileage. Under the present laws, HEW can test only
prototype cars which are "substantially the same kind of vehicle
that is produced" (and it should be noted that more than half,
or 55 million of the nation's cars were built with no pollution
controls required.) HEW officials make no bones about noting
that the prototypes are meticulously honed to perfection before
being tested and even then they sometimes have to take the
test more than once. Pending legislation would allow federal
inspection in the factory or on the showroom floor.

The Federal Strategy

The 1967 Clean Air Act set forth the basic strategy of the federal government's campaign against air pollution. It was enacted after it was obvious that previous admonitions were doomed. Only fourteen states had until then even begun to set their own standards and fewer than one hundred local governments had air programs. Failure stemmed from the inability of smaller jurisdictions to act against powerful industries who boosted the regional economy at the expense of befouling the air. The new act granted the federal government the power of an injunction to stop pollution if it presented an immediate threat to health. This meant that during a crisis like New York's 1966 Thanksgiving inversion, the government could order cars banned from the city, stop incinerators and factory combustion, and even order cutbacks in electric power generation if it were thought necessary to clean the air. Federal intervention has never been attempted, but cities like Los Angeles and St. Louis have taken such measures on their own, to one extent or another, during air pollution alerts and smog crises.

The main object of the 1967 act was to enable the federal government to gain control of air quality nationwide through the establishment by 1970 of regional air quality control districts corresponding to natural airsheds in urban centers. HEW was also empowered to issue air quality criteria to which the states would be given six months to respond by holding hearings and submitting their own standards to be approved by HEW. Then the states had another six months to set up plans for implementing the ambient air standards.

This all sounds simple enough but in practice it has worked abysmally. HEW lagged in setting up air quality regions and has fallen far behind schedule in issuing criteria and the accompany documents providing technical advice on control alternatives.

Moreover, the standards are applied to the ambient air of a district, leaving it up to the local jurisdiction to trace high concentrations of pollutants to their sources. If HEW takes action, the administrative delays, legal snags, and other loopholes in the law's procedures for enforcement can prevent justice for years. One classic case is that of the Bishop Processing Co., a chicken rendering plant in Bishop, Maryland, that nauseated nearby residents with an odor one described as "the smell of dead bodies" for fifteen years. Locally, attempts to shut down the plant failed from 1956 on, although a court injunction was issued against the owner, Harold Polin. The federal government intervened under the Clean Air Act of 1963, holding an abatement conference in 1965, and Polin was ordered to cease operation unless he installed odor controls. Nothing happened, so the case went to a federal district court, which ruled again that the plant must clean up or close up. Once more Polin was defiant and early in 1970 he was back in court. At last on May 21, 1970, seven years after the first federal action was taken, the U.S. Supreme Court closed the plant by refusing to take Polin's appeal.

The Clean Air Act has inherent weaknesses. The air quality districts cut across state lines to properly encompass the limits of the problem but there is no strong governing jurisdiction as a result. The districts, furthermore, are in urban areas while many rural towns now suffer from air pollution and many more, in the path of growth, will. When a state does try to enforce the standards, what polluter should it crack down on first? That question has not been answered because no state has yet presented HEW with a plan to regulate combustion fuels and smokestack emissions company by company so that ambient air levels will be maintained.

Setting Standards and
Public Involvement

Nevertheless, until pending proposals for national emission standards become law and air pollutants are regulated right at their source before they take flight, there is much that a citizen can do. Even though the Pennsylvania Breathers, mentioned earlier, have not yet fully attained their goals, they have turned the spotlight on air pollution and those responsible for it by participating intelligently as well as vociferously in the hearings on standards. One merit of the 1967 act is that it *does* encourage public involvement, and the education and information people at the National Air Pollution Control Administration under HEW have been most willing to provide data that would support the toughest possible air quality laws. Leighton Price, chief of information, and Sheldon Samuels, head of the Community Support Program, have been tireless in their efforts to contact citizens to discuss specific air pollution hazards and what measures are necessary to cope with them. The Community Support Program works closely with the National Tuberculosis and Respiratory Disease Association, United Cancer Council-Michigan Cancer Foundation, the League of Women Voters, the Conservation Foundation, and other groups, holding air pollution workshops around the country.

The proposed Virginia criteria—for particulates, 80 micrograms and for sulfur oxides, an annual geometric mean of .05 ppm—elicited the same kind of angry response that moved the Pennsylvanians. As a result of citizen testimony and publicized, informed protest, the standards that the state submitted to HEW were upgraded to 60 micrograms and .02 ppm

Citizens wrote nearly two thousand letters, then filled the hearing room, when the Illinois Air Pollution Control Board discussed standards for the Chicago Air Quality region in Aug-

ust, 1970. Over industry protestation that a 1974 cleanup deadline could not be met, the public persuaded the board to tighten
standards and advance the deadline to 1972. (It is also in Chicago that Mary Lee Leahy unsuccessfully filed suit against the
city claiming under the 14th Amendment of the Constitution
that she had been deprived of equal protection in obtaining
healthy air to breathe, because members of Chicago's air pollution committees and appeals board are representatives of polluting industries.)

Ultimately, the best air pollution law will contain national
emission standards exclusively or in combination with the current ambient air program. HEW submitted a report in March,
1970, on its investigation of the feasibility of such a system.
The agency recommended that the ambient air standards continue, with a significant change to allow public participation
in local hearings on implementation plans as well as on standards. The public is presently excluded from helping to draft
the enforcement plan, so this is the phase during which industry musters its lobbying experts to get in the last word. In
addition, HEW recommended national emission standards for
"stationary sources of air pollution." Thus the government
could force a factory that is planned or already under construction to include the latest pollution controls.

Another development that likely lies ahead is the so-called
"effluent charge," also discussed in connection with water. It
is a levy imposed on a polluter that is gauged according to the
value or *cost* of pollution coming from that source. The idea
is to set the charge high enough to force the polluter to clean
up or install a recycling system that will recover the wasted
pollutant by-products. Aaron J. Teller, dean of the school of
engineering and science at Cooper Union, expounded on such
an approach at a 1969 Congressional symposium. He noted that
while the U.S. consumed sixteen million tons of sulfur annually,
it emitted twelve million tons which could be called marginal
production because this tonnage cost more to recover than fresh

sulfur cost to mine. An effluent charge scaled to the value of that sulfur would force industries or utilities to develop sulfur recovery mechanisms or use fuels of far lower sulfur content, said Teller. The scientist pointed out that a large apartment building owner might be charged the cost of property damage caused by sulfur dioxide and particulates coming from his furnace and incinerator. The levy would in several years exceed the price of pollution controls or cleaner fuel, so it would pay the apartment building owner to conform right away. He could not pass too much of his costs along to tenants by raising rents, because of competition from apartment buildings far cleaner to begin with. In any event, the additional expenses all around would be small compared to the costs of pollution. At the moment HEW is studying the future application of effluent charges to cope with not only air pollutants but the staggering pile-up of industrial solid wastes, city garbage, and refuse.

One cannot sit back and wait for reform, however. Even if laws on the books do not give you much hope, there is enough for you to do to obtain better laws and flush out polluters who are taking advantage of you.

WHAT ELSE THE CITIZEN CAN DO

In late 1966, John Gardner, then Secretary of HEW, complained candidly that the fight for clean air was flagging badly. The reason was that "Air pollution, like any other environmental problems of metropolitanism, is compounded by improper land use, choked transportation, poor disposal of waste, and general lack of foresight in industrial and city planning." There are some indications that this picture may change soon, if it hasn't begun to already. In New York, Mayor Lindsay has been considering various measures to discourage automobiles in the city, including the drastic notion of banning all but commercial vehicles in large normally congested areas downtown. He has also called for higher gasoline taxes, a $10 annual fee on cars kept in the city and extension of the 6-percent sales tax to parking lots and garages. While the reason given for these proposals was the need for more city revenues, they might also have a salutary effect on automobile pollution.

New York City has also passed regulations requiring that the sulfur content of fuel be limited to one percent or less (v. the usual 2 or 3 percent), unless controls were installed on combustion units to produce equivalent emissions. These limits will undoubtedly be lowered further. New Jersey, the nation's most heavily industrialized state, passed such a law in 1968 and then reduced the sulfur limit to 0.3 percent, effective by 1971. New York State has ruled that the level be no more than 0.37 percent by 1971.

Certainly the utilities, the major contributors of sulfur dioxide, face a dilemma that will require far more effective planning and coordination to meet soaring demands for electricity. They will have to work closely with environmental agencies and conservation groups rather than being shortsighted and secretive as they have been in the past. They will also have to bring pressure on their suppliers, the coal and oil industries whose lobbyists have worked mightily in Washington and in state capitals to obstruct measures that would result in low sulfur fuels or cleaner alternatives.

Many natural resource economists refute the contentions of the coal industry that there is not enough low sulfur coal underground in this country, and the oil industry is most certainly not forced by the current system of tariffs, quotas, and protective allowances to market cleaner fuel oil. For one thing, it has been made more profitable for them to refine high-grade products such as auto and aviation gas than to produce a clean low-grade fuel oil.

As a citizen, you have access to state public utilities commissions and federal agencies (the Federal Power Commission and the Atomic Energy Commission) who grant electric companies permission to construct and operate their facilities and regulate the rates to be charged. If your Congressman is not willing to press your demands that a particular utility be more responsible with respect to both air and water pollution, then you should contact local conservation groups, or form one, and generate a public outcry similar to that which has so far been successful in preserving a scenic section of the Hudson River. (See Chapter 10.) Citizens' groups are an effective instrument to force utilities to break out their long-term plans and give an account of antipollution measures being taken. Consolidated Edison broke precedent to present a full picture of its future in August, 1969, entirely in response to citizen concern.

All the available evidence indicates that clean air will not be inordinately expensive and that, for example, less than one percent will be added to your electric bill to pay for necessary antipollution controls or cleaner combustion in power generation. But you are going to have to ferret this sort of documentation out of the education and information people at NAPCA, from local clean air organizations which have sprung up, and from outspoken health officials and economists who are cited in independent or government Air Quality studies (see Bibliography). You will have to find out when hearings are set for standards in your region and, if they have been held already, then begin to question the officials responsible for implementing the standards and see they are not relenting under pressure from industry. Has your state given proper notice concerning hearings, giving you time to assemble sufficient evidence to comment on proposed standards? Several states have changed their proposals at the last minute without giving the public notice to reprepare its position. Throughout the nation, city and state air quality boards have been packed with representatives of industrial polluters. Chicago is the most celebrated case. You should check out the credentials and interests of your local board and then register protest, if need be, by writing letters or organizing a citizens' workshop to galvanize interest in what is being done—or not done—locally about air pollution.

Many of the broad-interest conversation groups have participated in air quality hearings and they and clean air organizations (e.g., New York City's) have formed technical advisory committees to obtain health data and statistics pertinent to the alternatives in air pollution control.

HEW has issued pamphlets concerning criteria for sulfur oxides, particulates, carbon monoxide, hydrocarbons, and photochemical oxidants. These documents can be obtained from HEW, the Senate Air and Water Pollution Subcommittee, your congressman, and from state health or air quality officials. They are most useful for learning the health effects, other consequences and costs associated with various levels and types of air pollution.

How adequate are your local laws to cope with the problems outlined early in this chapter? Are there regulations forbidding the installation of new combustion equipment in houses or commercial buildings without prior inspection to see these furnaces meet strict air quality standards? Do planning agencies coordinate transportation systems, highway projections, industrial construction, and general land use to take into account the effects of planning on air? Are there local regulations on the use of fuels?

These are specific questions that one can ask now. Looking to the future, we are thrown back on John Gardner's probing thought. We are going to have to ask *ourselves* very basic general questions about the energy we demand for transportation, production of goods, and conveniences. How much of all this do we need? How much more are we willing to pay to obtain these benefits without being poisoned by their side-effects?

3 ❧ THAT AWFUL NOISE

NOISE CAN BE CALLED the "personal pollutant" because its psychological—*if not physiological*—impact is so varied. It is generally defined as unwanted sound, but people disagree widely as to what they want. One man's music is utter discord to his neighbor. Thus noise is the environmental distraction that undermines the right of "domestic tranquility" called for in the preamble of the U.S. Constitution. But alas it has not aroused the universal aversion or discomfort that has united so many people against smog and soot in the air, and pesticides and sewage in the water. Oscar Wilde may have seemed snide but he was not far off the mark when he wrote in *Impressions of America* in 1882, that "America is the noisiest country that ever existed. One is waked up in the morning not by the singing of the nightingale, but by the steel worker. It is surprising that the sound practical sense of Americans does not reduce this intolerable noise." The English author might be startled to find today that some Americans have muffled the street din by playing on their stereo sets "The Optimum Aviary," one side of an album titled *Environments* and featuring thirty-two birds in full song. (The other side is "The Psychologically Ultimate Seashore".)

There are those, however, who can't sleep or function to the sound of rustling leaves and songbirds. They are reassured instead by the background noises of the metropolis or a busy suburb—the tooting of horns, big trucks shifting gears, jackhammers and pneumatic drills, or lawnmowers, barking dogs, and braking delivery trucks. And the sounds that once bothered past generations—steamboat and train whistles, town criers, brass bands, and fire bells—now seem romantic compared to their replacements—diesels, sirens, and jets.

Exceeding all bounds, it is the thunderous roar of the jet and the prospect of the sonic boom that has finally aroused enough concern to make noise an environmental issue. *Nobody* wants to put up with subsonic jets overhead, and the prospect of having the house shake, windows crack, and ears ring from the boom of supersonic jets is quite frightening.

So the jet has forced us to consider the truly insidious effects of noise pollution, whether or not we like the sound. But in fact, at a level exceeded all or most of the time in cities and factories and even frequently in our homes, noise has been for some time a very definite hazard to the hearing system and health in general. "Prolonged exposure to intense noise produces permanent hearing loss," said a federal task force report, *Noise—Sound Without Value*, two years ago, adding that "Increasing numbers of competent investigators believe that such exposure may adversely affect other organic, sensory, and physiologic functions of the human body." Recent studies have been more definite. They have shown that noise at intervals or long duration, *besides* bringing on gradual deafness, damages the heart and vision, produces indigestion and stomach ulcers, builds up hypertension, and causes mental disorder that sometimes leads to suicide. Dr. Leo L. Beranek, a well-known acoustical consultant, has found that 25 percent of those at the age of sixty-five who have been exposed all their working days to a noise level of 100 dbA (a textile loom or riveting machine) suffer from serious loss of hearing. Tests in which animals are assailed with noise at levels not far above those many people are routinely subjected to produce horrifying results within a short time—cannibalism, homosexuality, loss of fertility, and outright heart failure.

Over sixty years ago, Nobel-winning bacteriologist Dr.

Robert Koch predicted "The day will come when man will have to fight merciless noise as the worst enemy of his health." That moment may be upon us. The World Health Organization estimates that annual losses in efficiency and occupational sickness from U.S. industries alone exceed four billion dollars. Up to sixteen million Americans are estimated to be working under literally deafening conditions. And the Veterans Administration now spends around $65 million annually to rehabilitate ninety thousand war veterans with deafness or disorders caused by the sounds of weaponry.

NOISE MEASUREMENTS

Thus, pleasing or not, noise is an environmental hazard to be ranked along with air and water pollution. It not only affects people emotionally, but it damages the ear and has many complicated effects on other bodily functions. How do you measure noise and at what levels is it hazardous?

Noise is described in terms of decibels, which define sound pressure on the ears. It is a point of confusion, however, that the decibel scale moves upward logarithmically so that only a slight increase, say of ten decibels, at a high level is actually a tremendous jump, perhaps a thousandfold or more. In addition, the decibel scale does not take into account shifts in frequency that also affect hearing. Various formulas have thus been worked out to obtain measurements that most closely correlate sounds with human responses. When you see measurements applied to sounds, always check to determine what scale is being used or how it is being weighted. The "A" Scale (dbA) is the weighting system that most acousticians feel approximates tonal effects on the hearing system by giving less weight to low frequency sounds. Other common variations include PNcb (Per-

TABLE OF SOUND LEVELS (dBA)

OVERALL LEVEL	INDUSTRIAL AND MILITARY	COMMUNITY OR OUTDOOR	HOME OR "INSIDE"
130 Uncomfortably	Armored personnel carrier—123		
120 loud	Oxygen torch—121		
	Scraper-loader—117		
	Compactor—116		
110	Riveting machine—110		Rock-n-roll band—108–114
100 Very loud	Textile loom—106	Jet flyover at 1,000 ft—103	
	Electric furnace area—100	Power mower—96	Subway at 35 mph—95
90	Farm tractor—98	Compressor at 20 ft—94	
	Newspaper press—97	Rock drill at 100 ft—92	
		Motorcycles at 25 ft—90	Cockpit-light aircraft—90
80 Moderately	Cockpit-prop aircraft—88	Propeller aircraft flyover at 1,000 ft—88	Food blender—88
70 loud	Milling machine—85	Diesel truck, 40 mph at 50 ft—84	Garbage disposal—80
	Cotton spinning—83	Diesel train, 40–50 mph at 100 ft—83	Clothes washer—78
60	Lathe—81		Living room music—76
	Tabulating—80	Passenger car, 65 mph at 25 ft—77	Dishwasher—75
			TV-audio—70
			Vacuum—70
50 Quiet		Near freeway and auto traffic—64	Conversation—60
40		Air conditioner at 20 ft—60	Bedroom air conditioner—55
30 Very quiet		Large transformer at 200 ft—53	
20		Light traffic at 100 ft—50	Leaves rustling—20
10 Just audible			
0 Threshold of hearing			

Note: Except where noted, these sound levels were measured at a typical distance away from the source.

ceived noise), ENdb (Effective perceived noise) and frequently
you see sound described in plain decibels with no weighting at
all. Damage to the ear begins at 85 dbA at frequent exposure.
This is the level considered dangerous at continuous exposure.
The table on page 67 of common noises and their dbA levels
is taken from material presented at a recent Senate hearing by
Dr. Alexander Cohen of the Department of Health, Education
and Welfare.

How do noises combine or interact? Do two jets flying side
by side overhead produce 206 dbA? The answer to that is no.
As to the effect of noises combined, there is a complicated
acoustical formula that, generally speaking, shows that two
noises only make a sound slightly greater than one. For example, two noises each at a level of 80 dbA would produce a
level of 83 dbA, and adding 75 dbA and 80 dbA would produce 81.2 dbA. What the "additive" formula cannot take into
account, of course, is the mental confusion created by a crescendo of different noises, particularly if they are not in harmony.

EFFORTS TO CURB NOISE

Legislatively as well as technologically, the U.S. lags far
behind other countries in setting standards and instituting controls to curb noise. There are countless local ordinances against
loud distractions—loudspeakers, unmuffled engines, stereo sets,
and radios outdoors, off-hours construction work, and nuisances
around hospitals and schools—but they are *seldom if ever enforced*. Only one city has successfully emulated the French edict
against car horns and that is Memphis, which imposed a ban as
long ago as 1930.

We like to compare ourselves to the Soviet Union's progress
in any number of areas, in rocketry and space orbiting, in in-

dustrial output and standard of living. It is disappointing to note, then, that the U.S.S.R. is ahead in a significant area of noise control, having for some time imposed a factory restriction of 85 dbA. Moreover, fifteen other countries, going back to 1938, have had building codes and land use criteria based on noise requirements. The only limits on construction in this country are minimal guidelines applied to federally supported housing.

Jets

In the early sixties, public outcry against noise grew along with increasing jet traffic. Airport authorities everywhere were sued by citizens angry over the thunderous roar that had not only disrupted living but had lowered property values considerably. The courts have ruled differently on such damage claims. There is reason for hope but as yet no clear pattern has emerged to settle the future of airports in crowded areas and to guide airport planners. In 1966, incensed over increased traffic to and from Kennedy airport, the town of Hempstead, Long Island, set stringent noise levels for jet overflights. But a federal court ruled that the Hempstead ordinance was illegal because it conflicted with federal regulation of airline commerce and was against the "total social interest" of keeping the airport going. Yet in the same year, an Oregon court rejected the argument that broad social benefits outweighed landowners' rights.

The most unfortunate victims of jet noise so far are those living near Air Force bases, particularly those fields handling supersonic aircraft. From fiscal 1956 through fiscal 1967, more than 35,000 claims for nearly $20 million dollars were filed over sonic boom damage, and about one third were approved in that period. The Air Force has taken the attitude that such claims are unjustified and that people should pay the price of national security. (This is the same reason given by the Atomic Energy Commission for steadfastly refusing to explain or hold a moratorium on many of its testing programs that are thought to have adverse environmental side-effects.)

When it was found that sonic disturbances had damaged scenic natural rock formations and cracked open archaeological treasures (including ancient cliff dwellings) in the southwest, Stewart Udall, then Secretary of Interior, in 1967 put together a team of scientists to study sonic boom. Later this group recommended in vain that the SST program be held up until the noise problem was solved. And since then Dr. William A. Shurcliff, a physicist, has launched the Citizens League Against the Sonic Boom and written the *SST and Sonic Boom Handbook* (see Bibliography).

If President Lyndon B. Johnson was not thoroughly convinced that noise was a problem when he allowed yet another committee to study it in July, 1967, just over a month later he was vividly impressed. On a September Sunday afternoon, Johnson attended services at the Lincoln Memorial honoring the poet Carl Sandburg, who had died not long before. Jets landing at National Airport just across the river drowned out speakers and poetry readers, and before it was his turn to talk, Johnson had had enough noise. Angrily he ordered Udall, sitting next to him, to get the Federal Aviation Agency to stop the jets while the President spoke. It was done and the planes circled up river (perhaps over Udall's house.) "The crescendo of noise— whether it comes from truck or jackhammer, siren or airplane— is more than an irritating nuisance," Johnson said later. "It intrudes on privacy, shatters serenity, and can inflict pain."

Although the liability of the sonic boom did not persuade Johnson to hold up the SST contract, his administration did make three moves toward remedying the noise problem. It

supported and enacted a Congressional proposal empowering
the FAA to set noise standards for aircraft. It published a good
task force study, previously mentioned, *Noise—Sound Without
Value*. And it recommended that the Walsh Healy Public Contracts Act of 1938 be amended to limit noise levels in most
plants doing business with the government to 85 dbA. That
amendment was weakened by the Nixon administration, which
changed the restriction to 90 dbA and, deplorably, has never
enforced the law because no federal apparatus has been set up
to check out noise levels affecting some twenty-seven million
workers in seventy thousand plants. (The amendment excluded
millions of other workers in industries doing less than $10,000
each in government business.) Federal officials were aghast—
and at a loss—early this year when one large plant bothered to
ask what it should do to comply with the 90 dbA standard.

Equally disappointing have been the FAA's moves to cope
with noise from overflights and the prospects of the sonic boom.
Originally the first restrictions to be issued by the agency were
to cover the 747 jumbo jet but would not apply to the planes
already flying. The hope was that pressure could be brought
to bear on these jets as their engines needed replacement or
major overhaul. But when the FAA finally announced a level
of 102 to 108 decibels (and this compares to the present
operating range of 110 to 120), it was not to go into effect
for the 747 but for planes whose certificate applications were
filed before the agency had begun to consider noise regulations.
It is perfectly possible *now* to quiet down subsonic fleets. But
the total cost for "retrofitting" the existing U.S. jet fleets is
estimated at $2 billion ($1 million for a 707 and roughly half
that for a DC-8) and would undoubtedly be passed along in a
5 percent increase in plane fares. The National Aeronautics
and Space Administration is presently providing the aviation
industry with $50 million in contract subsidies to develop a

quieter engine, and engineers are confident jet noise can be reduced substantially by fan blade modifications and other relatively uncomplicated adjustments. But what will come of all this is uncertain since the FAA has been reluctant to get tough in regulating aircraft noise. For one thing, the agency has declined to present a candid environmental statement regarding the SST, as required under the newly enacted National Environmental Protection Act, section 102. (See Chapter 10 for the legal implications of this clause.) The FAA has never conceded that the version of the SST prototype now being built will present a noise hazard that cannot be controlled by present technology. While the English government has proposed laws against SST flights over land, U.S. Secretary of Transportation John Volpe has only given his *personal assurance* that the plane should not be allowed to fly over land where its supersonic effects could wreak havoc. This sort of pledge is likely to crumble under the pressures of economic necessity, if the SST ever flies.

President Nixon has recently reorganized the environmental agencies of the federal establishment, but lamentably he gave no real status to noise control and left the administration of aircraft noise requirements with the FAA, the agency whose *prime mission* is to promote aviation and national supremacy in plane technology, which is hardly a role that stresses noise or other antipollution limitations.

Government Action

The September, 1968, Johnson administration noise report was a most forthright and complete document and is well worth reading. Among other things it made these positive statements:

Immediate and serious attention must be given to the control of this mushrooming problem since the overall loudness of environmental noise is doubling every ten years in pace with our social and industrial progress. If the noise problem is allowed to go unchecked, the cost of alleviating it will be insurmountable.

Our ultimate goal should be the achievement of a desirable environment in which noise levels do not interfere with the health and well-being of man or adversely affect other values which he regards highly. *The federal government must play a major role in achieving this objective.* (Italics added.) The problem is a public concern, and its alleviation frequently will require actions that transcend political boundaries within the nation.

The message was crystal clear. State and local authorities cannot always deal effectively with unwanted sound because they do not govern all the sources. New York City, for example, in late 1969, issued the report of a mayoral task force noting that the city had little or no authority to act against motor vehicles, transit and port authority activities, and any federally licensed or controlled operation such as aviation. Moreover, if states are to set noise standards, they must be uniform so as not to provide unfair advantages in market competition. Thus some federal authority will be necessary.

Since the Johnson administration report, virtually nothing has been done about the national noise level despite the general concession that it is one environmental crisis against which a campaign could make good headway at a comparatively low cost. What is being done beneath the federal government level can be summed up in a few sentences. Apart from the Memphis ban on honking horns, few localities have passed, much less enforced, antinoise laws. Wisconsin, New York, California, and Oregon have limited abatement programs. California, for ex-

ample, has set a noise level of 88 dbA on freeways but it is
rarely enforced. New Orleans, Boston, Chicago, and Los Angeles are contemplating offshore airports that float or are built
on land-fill. (The latter might be a good way of accommodating
the refuse overload if water pollution is controlled.)

For the time being, citizen nuisance, compensation, and
property damage suits must carry the fight. At least the courts
seem inclined to take a sympathetic view although, as previously
noted, no clear patterns have as yet developed. The New York
Court of Appeals ruled that a Lake George resident was not
only to be compensated for land condemned for a freeway
but for the damages caused by the noisy traffic that would follow
and be an intrusion "into the enclaves which many people have
sought as surcease from the hustle and bustle of modern life."
The U.S. Supreme Court upheld a Florida decision that property
owners could collect damages suffered due to the noise of jets
using Jacksonville Municipal Airport. And according to research done by the Conservation Foundation of Washington
(see Bibliography), the Iowa Supreme Court ruled in the case
of a neighborhood disruption that "noises may be of such
character and intensity as to so unreasonably interfere with the
comfort and enjoyment of private property as to constitute a
nuisance, and in such cases, injury to health . . . need not be
shown."

Such suits can only increase in frequency as air traffic grows
and the land available for runway additions or new airports
becomes scarce. The FAA has experimented with traffic pattern
changes and rotation of runways to try to relieve some of the
racket over Kennedy and La Guardia airports. But citizens have
not been satisfied. In the summer of 1970, New York Attorney
General Louis J. Lefkowitz filed suit against fifty-eight domestic
and foreign airlines, demanding that they immediately install
noise abatement devices on their jets.

RECOMMENDING
NOISE LEVELS

High decibel noises, like the roar of the jets, are not the only levels that cause harm and should be regulated. Believe it or not, noises you do not notice can slowly deteriorate the microscopic hair cells that send sounds from your ears to your brain. In response to noise, small arteries constrict and thus pressure is put on the heart through an increase in the pulse rate. Dr. Jerome Lukas at the Stanford Research Institute has found that noises that did not awaken sleepers nonetheless had produced noticeable fatigue by the time they arose. You may think you have gotten used to a noise but your bodily system has not and over a period of time your muscles, nervous system, and heart are taking a strain while your listening acuity is being dulled.

A panel of experts assigned by New York Mayor Lindsay to study noise took this phenomenon into account when it recommended a level of 30 dbA for residential areas at night (the limit in West Germany and other countries for some time) and 40 dbA during the day. The city government was not entirely responsive but did pass the nation's first antinoise building code that set the noise penetration level at 45 dbA on all new residential buildings. The Lindsay administration has also paid heed to the suggestion that the city use its purchasing power in the cause. New York has ordered quieter garbage trucks and is experimenting with rubberized cans and plastic bag substitutes that do not make such an awful din during the early morning hours. (The federal government ought to follow suit by imposing noise restrictions on the 35,000 vehicles it purchases annually as well as in a great variety of public service contracts.) The New York study recommended that a level of 85 dbA be enforced in busy areas and that noises above the normal speaking level of 52 dbA be reduced as soon as possible.

THE COSTS

One fact about combating noise pollution gives ground for optimism. *Noise can be brought under control or eliminated if necessary within the limits of present technology and often at a cost that is not excessive, not to mention the obvious immeasurable benefits.* The New York committee stated unequivocally that "The argument that noise control is too expensive for business is no longer valid, if it ever was." The Citizens Advisory Committee on Environmental Quality reported to the President in 1970 that "Technology is no barrier. Buses and trucks can be made much quieter by the simple device of larger and more efficient mufflers. Even jackhammers, it has been demonstrated, can be made relatively silent."

Moreover, a solid framework already exists for both government and citizen action to cope with the problem although, sad to say, apathy in both quarters continues to persist. John M. Mecklin, a *Fortune* magazine editor, noted in a most comprehensive article (see Bibliography) that the federal government had failed to take advantage of the interstate commerce act to prohibit high noise levels in factories or to invoke the same principle on industrial capital machinery manufactured for use in other plants. Thus a good opportunity has been missed for building silence into the industrial-technological system of this country. States and localities for various reasons have been unable to abate their ambient noise and Mecklin concluded that "with a few exceptions, businessmen have been surprisingly slow to recognize that noise prevention can be marketed; for example in advertising for quiet apartments or noiseless kitchen equipment." He added that "There is also a major public relations consideration in the growing feeling among environmentalists that corporations have no more right to dump noise on communities than air and water pollutants."

LAND USE AND CONSTRUCTION

An effective and inexpensive way of cutting down noise is to take it into account in land use planning and new construction. When urban designers plot new highways, all too seldom do they choose routes to minimize sound effects on people. Heavy trucks are often directed off of freeways and diverted through residential neighborhoods.

If highways are sunk between earth embankments, traffic noise can be reduced up to fifteen decibels. Trees and shrubs can be used to absorb both the noise and carbon dioxide that come from cars. And sending traffic through tunnels enables a city or town to economize on land by making use of the air space overhead, provided of course that a proper acoustical shield is laid down.

The construction industry is a major urban noisemaker. Yet riveters, pneumatic drills, air compressors, pile drivers, and all kinds of excavation and demolition equipment can be quieted down by mufflers or replaced by alternatives, generally at additional costs of around 5 percent, although there are exceptions. New quiet large air compressors used in cities cost 25 percent more and silent garbage trucks up to 15 percent.

Not only do the builders make noise on the job but what they erect are acoustical nightmares, and ultimately the value of this property suffers. "The purchase of an acoustically inferior hotel, motel, office, or apartment building may be an exceedingly poor investment," said the federal government 1968 noise report. "High-tenant turnover may require unfortunate building owners to convert very desirable and expensive rental space into storage space to satisfy occupant complaints of the lack of privacy and noise control; to make acoustical modifications in their buildings; or to sell their property at a loss."

Flimsy walls, thin floors, and lack of insulation allow the sounds of electric razors, flushing toilets, knocking pipes, rushing air and water, stairway traffic, kitchen activity, and family squabbling to permeate apartment buildings. And the city offices that rise everywhere are no better. The 1968 federal report was scornful of the building industry's pride in its achievements, noting that "the fact remains that conventional building techniques have produced some of the noisiest buildings in existence." The study listed these causes of noise in buildings:

1. *Mechanization:* increasing use of noisy high-pressure heating, cooling, and plumbing systems, power plants, and automated domestic appliances. Progress in mechanization is outrunning advances in machinery noise control.

2. *Poor acoustical design:* open-space layout without regard to separating noisy areas from those requiring quiet or privacy.

3. *Light-frame construction:* increasing use of thin wall and floor constructions and hollow-core doors, which are poor noise barriers.

4. *Poor workmanship:* careless work by builders in sealing holes, cracks, and noise leaks and installing equipment.

5. *High-rise buildings:* greater concentration of families in smaller areas result in noisier indoor and outdoor environments and increasing interfamily annoyance.

6. *Higher costs of sound insulated construction:* the increased cost of constructing a sound insulated building might range

from 2 percent to 10 percent of the total cost of the building, depending on geographic area, labor market, and other economic factors. Builders and owners of buildings are in a highly competitive market; therefore they are reluctant to adopt new features which may result in higher building costs or jeopardize their competitive positions and profit margins.

7. *The lack of mandatory acoustical criteria and enforcement:* until acoustical criteria are made mandatory and enforced by law, builders will continue to ignore them.

How expensive will it be to remedy the problems listed above? It depends where you live, but the expense is worthwhile. It cost me less than an additional one percent of the total cost to have my house designed and insulated to reduce noise, although I live in the country and the only close-by disruptions are the jets from Dulles Airport passing overhead. To completely soundproof a building in the city, using acoustical tiles and heavier frames, paying attention to design factors, and avoiding the use of small air ducts that are noisy because they take air under pressure at high speeds, the additional cost is estimated to be between two and ten percent, as noted above. Until the U.S. has to impose a national building code by default, it would pay for local governments to consult the research cited in the Bibliography of this chapter and obtain material on soundproofing from the Department of Housing and Urban Development and the American Institute of Architects, as the basis of drafting their own building regulations to prevent noise.

ON THE JOB

It cannot be emphasized enough that the psychological annoyance of unwanted sound effects on the mind are usually accompanied by physiological damage or impairment of performance. A Colgate University study found that industrial workers unconsciously spent one fifth of their energy fighting noise. And audible distractions are frequently cited as one cause of factory accidents. From the standpoint of health, occupational noise deserves much more attention and yet no national exposure guidelines have been considered. We turn to foreign data to note the worst effects on workers: abnormal heart rhythm and cardiovascular irregularities, hyperactive reflexes, insomnia, perpetual fatigue, and even sexual impotence.

If factory employees took full advantage of provisions granting them compensation for hearing losses, U.S. industries would act in a hurry. One estimate is that if only 10 percent of those eligible for compensation—the most obvious cases—filed claims, the cost in total damages would run up to a half billion dollars. This presumes that the average award is around $1,000 whereas in fact such damages have run around $2,000 a settlement. Fortunately for industry less than one thousand claims are now filed annually.

If noise prevention measures annually cost about $2 a worker —one federal estimate—it would cost up to $32 million to correct the conditions that plague the hearing of the 16 million workers cited in the beginning of this chapter. But this is an uncertain estimate and the costs of noise control in old or existing facilities that already lack proper health and safety features would undoubtedly run higher. In general, though, if included in new plant construction, noise control adds no more than 5 percent to the total cost of development. This may seem like a lot, but it really isn't when you consider the health benefits that result as well as increased production due to an atmosphere in which workers can concentrate instead of being continually distracted. A recent study by the National Bureau of Standards has revealed that factories and manufacturing plants are not the only noisy working areas to worry about. Computer and data-processing centers were found to present a very real danger to hearing. And most downtown offices, if they are not dangerously noisy, have little privacy from disturbing sounds.

CITIZEN ACTION

There are indications that citizen concern over the decibel din may at last be developing. The crusader who deserves the most credit is Robert Baron, who was driven to distraction by air compressors outside his Lincoln Towers apartment in New York. He gave up managing plays on Broadway and went full time into antinoise lobbying by founding Citizens for a Quieter City, Inc., and investigating what was being done or what could be done to bring peace and quiet to the U.S. after the example of European countries (who as yet tolerate a good deal of noise from motorbikes and other distractions). Baron instigated and served on the Lindsay noise committee. While he feels strongly that citizens' centers are able to arouse a good deal of concern and pull together helpful information, he is not sanguine about immediate reform. "There isn't a city code or state law that begins to approach the noise problem," he told me in the summer of 1970.

Thus it would appear that citizens will have to take it upon themselves to make their surroundings quieter. This they can do by participating in local planning decisions or influencing their town and city officials. Already in cities throughout the country, action groups have cited noise as a factor in protesting the decisions of federal and state highway planners. Noise suits and compensation claims are certainly a potent threat. And surely the construction industry by now should be wise to the unfortunate results of using thin, light materials and shortcut methods in the buildings they put up.

A full-page advertisement in the New York Times, March 3,

1970, by the Friends of The Earth, appealed to people's worst
fears about the sonic boom and noise in general as the "price
of progress." The headline, in bold half-inch type, noted that
an SST "breaks windows, cracks walls, stampedes cattle, and
will hasten the end of the American wilderness." That should
indeed stir a responsive chord, for you cannot help but feel
hemmed in when there is hardly a spot on earth where you can
stand even now and not hear aircraft at some time in the day.
Yet it was only a century ago when Henry Thoreau marveled
at the silence and solitude of a dense spruce forest near Mil-
linocket in the coastal state of Maine and exulted in his journal,
"Have we even so much as discovered and settled the shores?"

4 ✻ THE DILEMMA
OF GROWTH

FAIRFAX COUNTY, Virginia, is one of the fastest growing metropolitan suburbs in the nation. The towns of McLean, Falls Church, Vienna, and Great Falls are within easy commuting range of Washington. Yet until just over a decade ago, the region was mostly farm country rolling back from the flood plains and palisades of the Potomac through undisturbed marshes, rich meadows, and tall stands of oak, beech, walnut, and poplar. It is an area full of history, of fine old houses, a battleground where the Blue and the Gray surged back and forth during the Civil War when the country was a sort of no-man's land between North and South. Along some back country lanes at night one can still imagine the cavalry of Stuart, Mosbey, Early, and Sheridan galloping through.

It is sad that no one in authority seems to have considered the consequences of growth in this region. As a result, development has been aimless, unplanned, and disorderly almost beyond repair. Shopping centers and industrial parks went up where land was a bargain, not where they might have been convenient or followed a master plan. Houses were jammed together, the land around them stripped of vegetation to facilitate the earthmovers, so that soil erosion and sewage runoff plagues the Potomac River, not to mention the millions of gallons of raw or barely treated sewage that flow from three overloaded treatment plants. There was never any hope for the farmer. He was taxed out of business as his land was assessed at its fair market value—what it would be worth developed commercially or residentially—and thus metropolitan Washington was deprived of the meat and fresh dairy and vegetable products that came from this first-class agricultural land. There are few farmers left in Fairfax County, and now the wave of land speculation is squeezing out farmers in Loudon County on the margins of the Shenandoah Valley.

From a surprising quarter, a whistle blew to halt this growth —for a brief moment at least—in the spring of 1970. The Virginia Water Quality Board abruptly told developers in seventy-four of Fairfax County's 408 square miles that they could not hook up to the overloaded municipal sewage systems that were dumping raw effluent into the Potomac. Three large contractors promptly filed suit, claiming that the state had no right in effect to stop construction already underway or planned on houses valued at over $3 million. Other contractors joined in the protest, contending that 3,281 homes worth over $70 million would be held up without sewage facilities. The state countered with the admonition that until the people were willing to pay for additional waste treatment facilities to accommodate Fairfax County's rapid growth, the so-called "sewer ban" would stick.

The state's position may not be maintained through litigation, since the builders have a point when they say local authorities had issued permits. As a new policy though, the tactic is eminently proper. Maryland authorities across the river in Montgomery County have applied the same edict. The rule is simply that growth cannot proceed if it impairs environmental quality or results in pollution that will have adverse effects on downstream communities. "The sewer ban," says David Dominick, Commissioner of the Federal Water Quality Administration, "is a most effective way of bringing to people's attention the dilemma of growth throughout this country, the consequences of inadequate planning and failing to look ahead to the environmental side-effects of planning."

Fairfax and Montgomery counties' crises undoubtedly apply

to thousands of other growing regions. In fact, as this chapter was being written, the state of New Jersey imposed a ban on the Passaic Valley sewer authority whose aged and inadequate treatment plants have created a major crisis for 1.3 million people and hundreds of industries. Such problems may not invariably be solved by technology either. It is likely that, in many cases, growth has already used an environment to its limits, considering the patterns of land use followed in new developments and the staggering waste load of American families. Thus one should stop to consider the United States population problem, which is tied inextricably to the question of how much—or what kind—of expansion we can tolerate. When does progress become a case of going backward? When does growth run up against diminishing returns? Voices of concern over these questions have risen in both liberal and conservative quarters in the U.S.

MEASURING GROWTH

There is a growing feeling, particularly among young people, that the way we measure growth and progress is false because it does not take into account the values of our living experience —of peace, quiet, and privacy, the enjoyment of our surroundings, or the costs imposed by activities which deteriorate our environment. Thus the Gross National Product, the yardstick that measures the nation's economic health, is called by some "Gross National Pollution" because it is so inflated by the costs of correcting what are also counted as benefits, such as the automobile, industrial effluent, and the disposal of products made in the plants, from paper napkins and sheets to no-return bottles and cans. Economist Kenneth Boulding, a distinguished expositor of this point of view, noted in a speech during Earth

Week in 1970 that "The GNP is useful for the crude purpose of saying what goes up and what goes down, but it certainly needs to be replaced by a more accurate index which will take care of the bads as well as the goods. . . . When we pollute something, somebody else has to clean it up. That increases the GNP."

No less than Arthur Burns, former counselor to President Nixon and now Chairman of the Federal Reserve, took Boulding's view during hearings by the Joint Economic Committee of Congress in February, 1970. He conceded that the underlying premises of growth in the U.S. are based on false notions. "I look forward to the day," he said, "when our statisticians— when they cast up the national income accounts—take account of depreciation in our environment in addition to depreciation of plant and equipment." And he added that "Once they learn how to do that we will discover that the gross national product —which has been deceiving us all along since the figure we should have been looking at is the net national product—is a good deal lower than we think it is."

The perturbations of these two reputable economists are shared by more than young people decrying growth. In the top ranks of the industrial and business establishment of the U.S. there is uneasiness that the allied forces of economic and demographic growth have gotten out of control and are undermining domestic tranquility and the future health of the nation. In other words, we have arrived at what economists call the "diseconomies of scale," the limit at which it becomes more economical to supply less goods to less people than to pay for marketing inefficiencies caused by a congested system "Isn't it time to examine the growth syndrome?" asked Charles Luce, Chairman of New York's Con Ed, during a lecture to Yale students. "Surely we must reexamine population growth. I just do not see how in the electric industry—or in our economy as a whole—we can go on with the ever-increasing population and

keep this earth habitable for future generations into the indefinite future." And the chief of the world's largest utility
(by most indices) did not hesitate to say that "Economic growth
may be outmoded, may even be a dangerous national goal
twenty and thirty years from now."

The U.S. standard of living is unequaled. There seems to
be plenty of food and land for years to come. Congestion and
famine are well-known problems in other parts of the world
but why, ask many Americans, are population—and economic
growth that multiplies the effect of people—crises in this country? Isn't concern over population explosion really just a di-
version, an excuse for what we are failing to do to make our
cities governable and more efficiently distribute our food and
housing? And if we stop having babies, won't the average age
of Americans go up and at some point along the line won't the
nation be deprived of a generation of young leaders? Why
wouldn't it be far more productive to continue the present
economic practices that in theory allow anyone to make good,
and then redistribute our population in new towns and cities
to accommodate economic growth? Aren't mothers what God
intended women to be? Didn't the Bible tell us to bear and
multiply?

These are questions that bother people on both sides of the
growth argument. Too often the controversy is so sharpened by
ethical, social, and religious considerations that there is no
rational middle ground in which participants can come close to
agreement—at least over the main principles and facts that are
inescapable. The controversy thus boils at two extremes. One
side says you can't stop progress and accuses the other of being
Luddites—after those bands of antitechnologists who smashed
factory machinery during England's industrial revolution. And
the other side accuses its critics of being profiteers of the capi-
talistic system, unmindful of the pollution that results from
progress and unable to give up a good thing—the 10 percent

annual increase in revenues that stockholders now routinely demand of their corporations. Or, say the so-called Luddites, you are tied to religious or social beliefs that are simply out of date and inadequate in the face of today's problems.

I make no bones about where I stand—emotionally closest to the Luddites, pragmatically nearer the middle but agreeing entirely with Boulding and Burns that economic growth must be redefined and with Luce that indeed the *United States* has a *population* problem. Able demographers like Ben Wattenberg may feel that the "population explosion" is a scare tactic (a view he espoused in a *New Republic* magazine article) but the facts show otherwise.

What, it is then asked, are the relevant measurements of growth? In what areas can it be said that the U.S. has a problem? Since population is the taproot of economic growth, in what terms do we decide whether the American people must be limited in number? The yardstick is whether there is enough food, resources, and land to support our needs if we are not to continue gobbling up most of the world pie. Finally, but of no mean weight in this assessment, the American population must be judged in terms of a prevalent mood and theme in our history—that there is always room to stretch out, wander at will in the great outdoors, and get away from it all. More will be said about this in Chapter 7 on The Land, but it is noted here that a constant lure in American life has been the call of the wilderness. To be sure, a growing number of citizens in cities have been deprived of the experience of nature and perhaps they will contend that the "life style" they seek won't be enhanced by tall forests and great open spaces. I disagree. Experience shows that high-density living—deprivation of natural amenities—affects Americans more than any other people because the need for open space is so deeply ingrained in our character.

FOOD SUPPLIES AND LAND

In a speech during 1968, Dr. George A. Brogstrom, Michigan State food scientist and geographer, noted that "If all the food in the world were equally distributed and each human received identical quantities, we would *all* be malnourished. If the entire world's food supply were parcelled out at the U.S. dietary level," he added, "it would feed only about one third of the human race."

The fact is that while most of the people, or 400 million, in well-fed nations of Europe, the U.S., and the Soviet Union are glutted, more than two billion people in the world are either malnourished (meaning they are stuffed with foods inadequate in protein and essential vitamins) or undernourished (meaning they are not getting enough of anything). Thus the birth of one American equals the effect of fifty births in an underdeveloped country because the average American in his or her lifetime will consume such a staggering volume of food (not to mention other resources). On the average, in a lifetime each of us will devour 6,989 pounds of beef and game, 4,164 pounds of poultry and fish, 4,304 pounds of pork, 25,446 pounds of dairy products, 14,083 pounds of vegetables, and 9,043 pounds of fruit. Each of us requires about ten acres annually for life support—one and a half for food alone and three and a half for grazing the livestock that give food.

The so-called "Green Revolution," the new science of agriculture that multiplies food yields enormously, is a misleading hope. To begin with, in this country, first-class agricultural land is being gobbled up by urban expansion and general land development (five million acres annually) while at the same time harvest projections assume that current food acreage will remain constant. Citrus production in both California and Florida is threatened not only by pollution, as noted elsewhere in this book, but by commercial and residential construction. Jean Mayer, Harvard nutritionist who became President Nixon's food specialist, feels very strongly that we may have to decrease our population in order to stay even in food production. "I am afraid," he told a Congressional panel, "that as we go along, a greater and greater proportion of our GNP is going to be used for corrective measures in situations which might not have arisen in the first place if our social planning had been somewhat better than it is."

Moreover, ecologists have discovered dangerous worldwide side-effects from U.S. sponsored agricultural projects. Irrigation systems have poured salts, nitrates, herbacides, and pesticides into river systems and into the earth in amounts that produce adverse effects on health and eventually doom the productivity of the soil itself. Nitrate levels in the water tables of some areas of the U.S. have already caused alarm.

Dr. Preston Cloud, Jr., biologist and Chairman of the National Academy of Sciences Committee on "Resources and Man," gave a disheartening report to a Congressional committee holding population hearings in the fall of 1969 (see Bibliography.) But first he had this to say:

I define an optimum population as one that is large enough to provide the diversity, leisure, and substance whereby the creative genius of man can focus on the satisfactory management of his ecosystem, but not so large as to strain the capabilities of the earth to provide an adequate diet, industrial raw materials, pure air and water, and attractive living and recreational space for all men everywhere into the indefinite future.

In terms of food supplies alone—skipping other essential

resources—Cloud said the current U.S. population of over 200 million and the world population of 3.5 billion are beyond the optimum. You could increase food supplies nine times by the "Green Revolution," Cloud conceded, but you would still have to assume a staggering rate of malnutrition in order to accommodate the thirty billion people that present growth rates will produce worldwide by the year 2075. In his opinion the projected world population of seven billion (300 million in the U.S.) by the end of this century is about all the world can sustain. And he noted that without unforeseen increases in natural resource yields, today's standard of living in the U S., let alone the rest of the world, would be impossible to maintain. How about food and minerals from the sea? That is a misplaced hope like the others, it was noted in a report submitted by Cloud. Even if the oceans can be kept uncontaminated by chemicals and oil, and while the marine harvest could increase two and a half times, only the land can provide foods of adequate calorie content. As for minerals, said the NAS report, "there is little basis for assuming that many marine and chemical resources are of large usable volume or feasible recoverability or that for many essential substances there are any resources at all. . . ."

NATURAL RESOURCES AND WASTE

Roger Revelle, Director of Harvard's Center for Population Studies, has noted that *only* if the U.S. population had *actually decreased* from 133 million to 67 million between 1940 and 1965, would we have held the line on our consumption of natural resources. He arrived at this conclusion after studying the U.S. Statistical Abstract and finding that while the popula-

tion grew 47 percent in that period, our GNP tripled and
American per capita production or use of materials zoomed
upwards. Annual paper consumption went from 250 to 510
pounds; gasoline from 750 to 1,500 pounds; coal and oil from
four-tenths of a ton to one and one-third tons; and American
farmers used thirteen times as much nitrogen fertilizer—5.3
million tons compared with 420,000 tons in 1940. New materials increased the GNP "throughput" tremendously. For example, per capita production of plastics jumped from thirty-
four to sixty pounds between 1960 and 1965.

It is variously estimated that between 50 and 70 percent of
the world's essential resources—fuel for electrical power and
transportation, basic as well as exotic metals, many different
chemicals, and other materials such as rubber—go to support
the U.S. population, or 6 percent of the people in the world.
At the same time, the U.S. holds out to the other nations the
hope of attaining our standards of living. "And isn't that only
right and equitable?" Stewart Udall, former Secretary of the
Interior, once mused. And then he asked, "What if all the other
countries—the underdeveloped countries—decided to hold on
to their resources so they could achieve that higher standard,
generate more electricity, build roads and have more cars to
run on gasoline, develop new industries, and so forth?" The
thought of this happening upsets all the nice equations of resource experts and it makes demographers and political scientists
shudder.

"It is not too comfortable for a country like ours to have an
optimistic outlook for itself and for most of the world to have
a much more pessimistic and uncertain outlook," Joseph
Fisher, President of Resources for The Future, told the Congressional population panel. Even in this country our inability
to conserve resources or use them efficiently takes a mounting
toll. As Fisher, a man respected for his moderate tone, put it:
"We are slowly dropping back and losing the war against air

and water pollution and against urban and rural land degradation."

THE QUALITY OF LIFE

This is a phrase that has suffered in rhetoric and overuse during the past year. It means different things to different people, anyway. It is the simplicity extolled by Henry Thoreau, who wrote, "I wanted to live deep and suck out all the marrow of life, to live so sturdily and Spartan-like as to put to rout all that was not life." And quality is also the abundance most Americans hope to obtain, so long as acquiring it does not produce pollution or monotonous ugliness across the landscape. It would be ideal if we were to achieve a state of quality nationwide that lay somewhere between Thoreau's lean style of living in the cabin he built beside Walden Pond and the plentitude that has come to symbolize the American way.

Technology alone cannot help us arrive at an answer. For the geniuses of industry and science have been of little help in making congested cities more liveable or air and highway travel any more efficient, mainly because of our ways and demands in convenience. And it doesn't seem likely that our habits will change overnight—if ever. The immediate answer is to hold down the number of people. It is in the attainment of quality that economic and demographic growth intertwine. Certainly one way of maintaining our standard of living, and a satisfying rate of consumption, would be to develop what economist Boulding calls the spaceship economy. This approach views the Earth as an Apollo moon capsule, in which all systems must interrelate precisely, there is no room for waste, and recycling is the order of the day. An error in efficiency, and the capsule is in deep trouble. In order to count as measures of prosperity or economic health, in Boulding's scheme, all industrial products have to be salvaged and reused. Otherwise they are gross national pollutants. "The essential measure of success of the economy is not production and consumption at all," he has written, "but the nature, extent, quality, and complexity of the total capital stock, including in this the state of the human bodies and minds included in the system." The implicit goals of the spaceship economy are survival with the least possible consumption, reducing or eliminating obsolescence, and recycling all waste.

However, even if patterns of food and material consumption are simplified and made far more efficient as well, the facts still lead to the conclusion that population growth must be halted. A number of government health studies indicate that insecurity, the individual's sense of hopelessness, is compounded by the number of people he must compete against for education and a job and place to live (not counting the frustration caused outside congested living areas by crowding on the highways or in the national parks). At Berkeley in 1965, students felt so overwhelmed by the size and depersonalization of the University of California that they imitated IBM cards and carried placards saying, "Do not mutilate, fold, or spindle." It had gotten to the point where the machine cards were getting more careful handling than the kids.

There is no question either but that the costs of development and technology are usually exponential to the rate of growth (are diseconomic). In other words, instead of services for a growing number of people being less costly per person (as in the bulk rate theory), they are more. Communities mistakenly encourage new development on the premise that it will increase the property tax base and thus give more revenues, but usually such growth costs more than it provides because roads, schools, and a host of other services must be improved, new sewer and water systems must be built, fire and police coverage expanded, and so forth. As *more* people crowd together *more*, these services become *more* expensive to obtain because of any number of complications. It may be that the fire department needs a newer kind of pump or ladder truck plus additional hydrants. Maybe the police have to build a holding cell or hire detectives. Or maybe as new demands are put on environmental necessities such as water and vegetation (open space), previously nonexistent costs of protecting these resources are introduced. They usually add up in subtle or intangible ways and are thus not counted or are misunderstood as arguments against growth.

THE SST

In a speech that he gave several times during 1969, Admiral Hyman Rickover, father of the nuclear submarine, warned that "Just because we can build an SST doesn't mean we should. Just because we can build a dam, or fill in a bay, or strip mine an entire country doesn't mean we should. Because," he added, "America is judged not only for the quantity of her factories but for the quality of her society as well."

Indeed, when it comes to questioning progress or priorities, the controversy over whether the U.S. should go ahead and develop a supersonic transport has taken on tremendous significance. Distinguished scientists and even the President's Environmental Quality Council Chairman, Russell Train, warned that the consequences of an SST could be dire. A fleet of SST's, he said, could eventually increase the upper atmospheric water content "by as much as 50 to 100 percent." And Dr. Gordon J. F. McDonald, a scientist on the Council, said that such an increase in water could alter weather patterns significantly by increasing world cloud cover. Worse yet, it could decrease the amount of ozone that shields the earth from lethal doses of

radiation. In addition, Train and others noted that the SST
would contribute to both air and noise pollution. Their Cas-
sandraic warnings so far have not convinced the White House
and the Department of Transportation that the costly project
should be halted until more is determined about its side-effects.
But the questioning of the growth-and-technology syndrome
was unprecedented. The House of Representatives narrowly
voted an appropriation of $290 million to continue develop-
ment of SST prototypes and, as this book was published, it
appeared that the Senate might block funds for the SST. Congressman Henry Reuss called the project "a mockery of needed
national priorities," and those agreeing, like the Sierra Club,
noted that it would be far more productive to use the $290
million to augment other budgets such as urban mass transportation ($204 million), air pollution control ($106 million),
and consumer health and protection ($85 million).

THINKING AHEAD

In a speech during Earth Week, ecologist Kenneth Watt
despaired over our inability as a nation to think long and hard
about the future effects of present decisions and policies. "We
live in the present and don't think much about the past and
don't think much about the future," he said. Indeed, American
society has become so mobile that few people today become
attached to one community all their lives. A series of profiles
in the *Wall Street Journal* in 1970 gave much attention to
the fact that too few active young American couples become
deeply involved in the problems of growth, land use, schooling,
and planning ahead in the communities where they are
"stationed."

Our mobility notwithstanding, the challenge of national

growth—demographic and economic—would seem to give us a new opportunity to lay new groundwork. What better beginnings could be made than to set forth the objectives of the spaceship economy—quality over quantity—here on earth instead of in the galaxies. Hopefully, such a commitment might galvanize world feelings on a number of issues that now divide it. For example, instead of consolidating its position as a nation whose majority are voracious consumers, the U.S. might set an example by applying the brakes to growth and becoming far less wasteful.

Zero Population Growth

The first goal should be a national zero population growth rate. ZPG is not only the name of an organization under Paul Ehrlich, the fervent crusader who wrote *The Population Bomb,* it is an objective recommended by an overwhelming majority of experts who have spoken or testified on the population problem. Thus it is none too soon to begin because even if *right now* we achieve a national birth rate that does not exceed the death rate, it will be seventy-five years before the population is stable—at a level of about 275 million—because of the maturation of the postwar babies (women 20 to 29) who will reach the main childbearing years late in this decade. It is too much to expect that these new young couples will not have children at all, much less the two per couple that will eventually result in stabilization.

ZPG is advocated by Congressmen such as Paul N. Mc-Closkey, Henry Reuss, and George Bush and by Senators including Joseph Tydings and Robert Packwood. Writing letters to these men and others urging that Congress declare that ZPG become national policy would help. But specifically, other measures must be instigated to make that policy possible. Through federal legislation—not only through court interpretations or state action—restrictions against abortion must be abolished and women be given the absolute right to decide whether or not they want to bear children. More than a dozen states have reformed their abortion codes, but only in New York have new laws not been unduly restrictive or complicated. Some states demand costly medical or psychiatric consultation, general hospital review, evidence that childbirth will cause an unbearable strain, or state residency. As usual, from evidence gathered to date, the poor and those who do not want, or cannot afford, more children have been unable to get abortions. In addition, the stigma against abortion, or the legal restrictions placed on it, have kept up the rate of criminal operations on pregnant women during which permanent damage or even death often is the result.

Certainly the present tax laws must be changed to encourage small families. Obviously such reform should not penalize those who already have more than two children, but it should be scaled as an incentive to hold down instead of increasing the number of children, as is now the case. Senator Packwood has introduced legislation limiting dependency deductions to two children ($750 per child but not applying to adoptions.) The bill would go into effect after 1972. It is currently being studied by the Treasury Department.

A national goal of no more than two children per family in order to eventually achieve a national ZPG would thus be enhanced by liberalized abortion rules and tax reform. ZPG should be the first pronouncement of a population commission, proposed by President Nixon and enacted by Congress. While the administration increased the budget for "family planning," this expression should be replaced by "birth reduction" and more aid should be provided in contraceptive methods. The term "family planning" implies that birth control is no more than a convenience for stretching out or scheduling childbirth. Contraception is currently the subject of controversy because birth control pills high in estrogen, according to English surveys, have caused fatal blood clotting. Pills low in estrogen are of course not so effective. Diaphragms, intrauterine devices, and vaginal foams are less effective. Male sterilization through a vasectomy—as done by Paul Erhlich and some of his associates —is judged to be a simple and harmless operation but it meets natural resistance. The desire for succession runs so strongly that few men want to foreclose their options to have children, should their families suffer tragedy or should they remarry. The answer is a combination of approaches—more federally backed research on contraception coupled with removal of restrictions on abortion. Eventually the government ought to provide abortion clinics as part of its medical services.

In calling for such measures, one gives the impression that unwanted children are the only problem. This is not the case, at least according to the most convincing evidence. Polls show that the desired or ideal family size is 3.5 children, meaning that many couples told interviewers they wanted four or more children. The social climate that promotes population expansion is deep-seated and involves many difficulties. Roger Revelle pointed out during the population hearings that unmarried people are looked down upon in this country. The woman is expected to start man-hunting and raise a family the minute she leaves school. "We will never persuade women to play their full role in society, let alone to have fewer children, unless we give them opportunities for something else to do, something meaningful and important," Revelle said. This would mean not only changing attitudes about the role of the woman— liberation, if you will—but concrete steps to give women greater educational and job opportunities. Along with this change in emphasis should come the realization that not only is late marriage or a small family more conducive to domestic happiness and the full development of children's potential, but perhaps not marrying or not having children at all is a worthy objective. This of course means that unmarried people will have to have sexual gratification in their lives without having to be furtive or hypocritical. Only enlightened pronouncements or action at the highest levels can begin such a social revolution, but the facts of growth nationally and worldwide demand nothing less. The times are not only changing, as Bob Dylan used to sing, they have changed. Institutions, the mores, people, have got to catch up.

As a parallel to reform in this country, U.S. foreign aid programs for birth control have got to be accelerated. William H. Draper, Jr., Chairman of the Population Crisis Committee, told the Senate Foreign Relations Committee in 1969 that the family planning appropriations should double immediately and that birth control should get the highest priority in the aid program. He could have added that agricultural assistance has too often boomeranged environmentally and that, anyway, the "Green Revolution" is only a temporary expedient to alleviate the starvation of a stable, not *growing* population.

Individual Action

Since both population and economic growth stem from individual behavior, the opportunities for action are great, although as indicated there are not nearly enough alternatives or incentives for putting a "hold" on the countdown of expansion (6.6 babies a minute in the U.S.) and a ballooning GNP. As this book was published there were at least a dozen bills proposed in the House and Senate that would proscribe population control measures, and several of these provided openings for considering the economic-environmental aspects of population.

Today we must be judged competent before being licensed to drive a car and shoot a gun, although a marriage license is

more of a formality. It could very well be—and I would like
to see the merits debated—that the marriage contract will include a provision concerning child rearing, or that Americans
will some day have to pay for a permit to have a child. Con-
stitutionally, this may seem shocking. Yet it is doubtful that
the founding fathers could foresee that the birth of a new
citizen might diminish the rights and opportunities of other
citizens. There would be a danger that certain groups in this
country might take advantage of such a measure on apartheid,
anti-Semitic, or other racist grounds. If such a control were instituted, it would have to be applied uniformly and only as a
penalty to excess growth. Neither should it be so inflexible as
to deprive poor people unnecessarily, although there are those
who argue that poor parents should be restricted in having
children if their poverty shackles development of a child.

A revolution in social policies is in order because, as Dr.
Judith Blake Davis, University of California at Berkeley demo-
grapher, testified:

We penalize homosexuals of both sexes. We insist that women must
bear unwanted children by depriving them of ready access to abortion, we bind individuals into marriages that they do not wish to
maintain, we force single and childless individuals to pay for the
education of other people's children, we make people with small
families support the schooling of those who have larger ones, and
we offer women few viable options to full-time careers as wives and
mothers except jobs that are, on the average, of low status and low
pay. In effect, we force a massive investment of human resources into
the reproductive sphere—far more than we need to invest.

Such are the costs of growth.

5 ⚘ GARBAGE AND ALL THAT TRASH

THE NATION'S staggering daily output of refuse presents us with both a crisis and an opportunity. This duality makes garbage perhaps the most challenging environmental emergency and certainly the most intriguing, because of the manner in which our ingenuity is being tested. No other problem can be so thoroughly dispatched by Technology with a capital T. And solutions ought to yield profits in the form of conserved and marketable natural resources. In fact, if we approach this crisis by using the system analysis techniques already applied to military and space programs, there should be precious little garbage left to dispose of. It is estimated that in urban wastes alone, we could salvage twenty-five million tons of paper, fourteen million tons of glass, twelve million tons of metals (not counting automobile scrap), 2.2 million tons of rubber, and an endless variety of other materials, including gold and silver.

It is discouraging to note, though, that we have only *begun* to deal with the garbage threat, recognizing that it is a major cause of unrest in cities, particularly in ghettoes, a key factor in land use planning, a major esthetic detriment, particularly along roads, and a nationwide source of air and water pollution. State and local governments have been waiting for the federal government to develop a strategy and set guidelines. And they may have to wait awhile. Because, as put succinctly by Leon Billings, Senator Edmund Muskie's staff man on the

Senate Air and Water Pollution Subcommittee, "We are in solid waste legislation today where we were with air pollution in 1963." And, as you may remember from previous chapters, air pollution programs are up to three years behind the battle to clean up waterways.

The U.S. Congress took notice of the garbage crisis only as an afterthought. In 1965 when the Air and Water Pollution Subcommittee was drafting new air quality legislation, the senators were stirred by publicity over the effects (on air) of burning in open dumps and incinerators. They agreed to tack on an amendment known as the Solid Waste Disposal Act of 1965.

Solid waste was defined thus:

. . . garbage, refuse, and other discarded solid materials, including solid waste materials resulting from industrial, commercial, and agricultural operations, and from community activities, but does not include solids or dissolved material in domestic sewage or other significant pollutants in water resources, such as silt, dissolved or suspended solids in industrial waste water effluents, dissolved materials in irrigation return flows, or other common water pollutants.

This definition is important primarily in that it has provided an excuse for various federal agencies to assume or avoid jurisdiction. Practically speaking, it has been stretched considerably to allow environmental planners to probe the interactions of all waste activities and to develop policies that would prevent consumer goods and industrial scraps from becoming "solid waste" in the first place. For example, flushing garbage down a kitchen sink grinder or burning refuse in a backyard incinerator lightens the burden of the local sanitation service but adds to water or air pollution. Then of course, once waste is collected, depending on how it is handled, a municipality is apt to create a big portion of air and water pollution. One team of scientists racks its brains for methods of treating the volume of discarded bottles, cans, old appliances, and even junked cars, while another team is preoccupied with a recycling or reuse scheme that will convert these wastes back into usable materials before they are ever taken to the dump. It is a complex and confusing business, often a vicious cycle, and a matter of increasing concern. Former Boston Mayor John Collins testified at a hearing on additional legislation in 1969 that "Clearly, if solid wastes are not properly treated and disposed of, they may undermine all of our other efforts to improve environmental quality." The Nixon administration's reorganization plan for government antipollution activities concentrates solid waste programs in the proposed Environmental Protection Agency. Up to now, even though the Congress acted in 1965 to establish a Bureau of Solid Waste Management to administer a research and development grant program, little of the present knowledge for coping with solid waste has been implemented. Alas, we the public seem bent on continuing our wasteful ways.

INCIDENTS

Nowhere has garbage had a more debilitating effect on human beings than in the congested city. In March, 1968, the National Advisory (Kerner) Commission on Civil Disorders concluded "that slum sanitation is a serious problem in the minds of the urban poor." Among the reasons for this, the report cited "higher population density; lack of well-managed buildings and adequate garbage services provided by landlords . . . high relocation rates of tenants and businesses, producing heavy volume of bulk refuse left on streets and in buildings; [and] different uses of the streets—as outdoor living rooms in summer, recreation areas—producing high visibility and sensitivity to garbage problems. . . ."

Slum sanitation is worse than ever In East Harlem one August Sunday afternoon in 1969, a gang of youths called the Young Lords burned garbage right in the middle of street intersections to protest the filthy, stinking environment that they said had been caused by inadequate garbage collection. Next day, some sixty demonstrators, including housewives pushing baby carriages, blocked streets in the Flatlands section of Brooklyn to publicize their outrage over hordes of rats that were not only rummaging through loose swill but terrorizing children. These uprisings continued into the summer of 1970, first in the Brownsville section of Brooklyn. Amidst the pervasive stench of decaying food in uncollected containers or spilled onto the streets, while sanitation equipment was out of service for repairs, the citizenry erupted. Youths ignited first the refuse, then vacant, dilapidated buildings, and finally they broke into and burned stores, setting loose a wave of vandalism that had to be quelled by the police. But what counted to the Brownsville rioters was that their fury had also attracted a squadron of sanitation trucks as well as a visit from the Mayor himself. One cannot condone this manner of protest, but one must be sympathetic to the cause and take note that such conditions ought to be given far greater priority in city planning and management in the U.S. today. New York is desperate to find ways of handling waste. Even when it is collected, it remains a monstrous problem. The city's twenty-four thousand tons of daily waste grows 4 percent annually and will fill up the present three landfill dumps by 1976. Construction of incinerators and new kinds of scrap recovery plants lag behind. As if this were not enough to stymie Mayor Lindsay's environmental officials, federal government scientists disclosed that the city's dumping of sewage sludge some twelve miles off Long Island has killed virtually all marine life over a twenty-square mile section of *ocean* bottom. New York is only one of the many large U.S. cities, and just about every one is fast running out of room and methods for handling wastes.

A presidential task force study entitled *A Comprehensive Assessment of Solid Waste Problems, Practices and Needs* (see Bibliography), commissioned by Lyndon Johnson in 1968, stated that "It is the urban solid wastes, generated at a rate of 256 million tons a year that poses the crisis with which the nation must cope immediately."

HABITS

In both city and country, the patterns of consumption described in the last chapter, coupled with our "affluent," sloppy habits and the lack of incentives or penalties to make the manufacturers of consumer goods responsible, have taken a high toll of environmental quality. Thus did Edmund Muskie, Chairman of the Senate Subcommittee on Air and Water Pollution, open a waste hearing with these despairing remarks:

We have made a beer can which will last for generations. We have developed plastics, containers, and wrappers which will last as long. We have learned to coat our paper products and chemically treat them so they degrade more slowly. At the same time we have developed and sold the concept of single use. There is no returnable or reusable beer can. Chemically treated paper and most aluminum and glass products are designed to discard. Economic substitutes are not available. As a nation we can take credit for eliminating return bottles and for achieving a point where the cost of repair of an obsolete appliance usually exceeds the value of the appliance, as my wife points out to me frequently. But as a nation we can no longer afford this attitude. We are a resource-scarce nation with a limited land and air supply.

Muskie's words were amplified by a witness, Robert Finch,

then Secretary of Health, Education and Welfare. In a tone just as pessimistic, he presented an assortment of grim statistics revealing just how wasteful we Americans have become. In 1968, we did indeed discard forty-eight billion cans and twenty-six billion bottles, virtually all unsalvaged, and 100 million automobile tires, of which only thirty million were reused. In 1920, our wastes averaged 2.8 pounds per person a day but had grown to 5.3 pounds in 1969 and will mount to eight pounds by the end of this decade.

Hospital wastes alone, Finch said, had risen from 3.8 pounds per person per day in 1955 to an average of nineteen in 1969. Under questioning, he admitted that this figure is probably doubled in ultramodern hospitals where practically nothing is saved—from syringes to sheets. And Finch might have added that it was the Public Health Service under HEW that has been urging hospitals to practice the one-time-use concept to avoid the risk of spreading infection *inside*. Finch said that fourteen thousand cases of persons being bitten by rats (a figure he thought was "grossly underestimated") and twenty-two human diseases, including encephalitis and hepatitis, were associated with unsanitary waste disposal (*outside*).

To provide an overall picture of the waste load, here is a table based on a survey of *Community Solid Waste Practices* by HEW during 1968. It covered a cross-section of cities whose populations are over five thousand and it counted only "material known or estimated to be collected." It did not include privately disposed refuse. So the figures presumably are low.

The 1968 presidential report mentioned earlier, which was chaired by Stanford University professor of environmental engineering Rolf Eliassen, noted tersely that "For many Americans, solid waste problems begin and end in several thirty-gallon containers outside the back door." And John C. Gridley, Chairman of Chemung County, N.Y., Board of Supervisors,

AVERAGE SOLID WASTE COLLECTED—*1968*
(pounds per person per day)

SOLID WASTES	URBAN	RURAL	NATIONAL
Household	1.26	0.72	1.14
Commercial	0.46	0.11	0.38
Combined	2.63	2.60	2.63
Industrial	0.65	0.37	0.59
Demolition, construction	0.23	0.02	0.18
Street and alley	0.11	0.03	0.09
Miscellaneous	0.38	0.08	0.31
TOTALS	5.72	3.93	5.32

told the Muskie subcommittee, "Garbage is a very unattractive subject and people want to forget about it as soon as it is out of sight. Too often it has been felt that if the local garbage collection and disposal system is cheap, it's good."

NINETEENTH CENTURY TECHNOLOGY

What is particularly galling in this nation of sophisticated scientific accomplishments is the fact that "cheap" waste systems are *not* inexpensive, and they employ—as Finch and others have testified—nineteenth century methods. In fact, about the only innovation is the use of trucks instead of horses to move the garbage and mechanized equipment to move and compress the refuse. And second only to tree-topping—a limited profession of the logging industry—sanitation work is the most dangerous occupation in terms of health and safety.

The 1968 HEW survey found that less than 6 percent of

twelve thousand dump sites met "sanitary landfill" requirements. In other words, the fill was not covered each day with
a layer of dirt to prevent litter from blowing around and attracting rodents and other pests or the refuse was dropped in a
ravine, freshwater marsh, or coastal estuary and thus threatened
drinking water supplies or polluted a vital marine environment.
Estimates range widely on the number of dumps that are not
so-called "sanitary landfill" projects but are open-burning sites
that flaunt air pollution codes. There may be 200,000 of these.
Of the three hundred municipal incinerators surveyed in 1968,
over 75 percent were inadequate and thus sources of air pollution. All told, HEW found that over 90 percent of 8,500 community waste collection systems failed in one way or another to
meet basic environmental standards. Russell Train, Chairman of
the White House Environmental Quality Council, has pointed
out that instead of providing an ideal solution by filling in old
quarries, gravel pits, or ugly strip mines, "All too often, in
search for dump sites, cities and industries settle on out-of-the-
way natural areas which have high values for educational and
recreational uses, or have irreplaceable historic or scientific
values."

While cities are just beginning to experiment with "twentieth century" alternatives, to date no American community has
devised a long-term, comprehensive waste disposal plan taking
into account land use factors, the potential of reusing refuse
scraps for fertilizer or raw material, and avoiding some form
of pollution or esthetic degradation. Of the $4.5 billion now
spent each year nationwide in sanitation control (a figure that
does not account for homeowners' and environmental costs),
over 75 percent goes into collection and transportation equipment. In general, waste is still approached as a treatment, not
a management problem. "We're worrying about more room for
dumps," says Leon Billings, "when we should be reducing the
load by enacting recycling legislation."

AN INVALUABLE
RESOURCE?

The preface of a 1969 National Academy of Sciences Study, *Policies for Solid Waste Management,* begins with a statement that is obvious but is generally forgotten:

Matter can neither be created nor destroyed. Man processes and uses matter. In so doing he may change its chemical form or alter its physical state; but in some combination of gases, liquids, or solids, all of the original material continues to be part of the world about us.

This is the reason that resource economists and mining engineers are so disturbed by current waste disposal practices. Some two thousand new industrial products annually are designed to become obsolete. Instead of being salvaged and reused, they end up as pollutants. Consider these facts. Eight to eleven percent of the materials in municipal waste are metal products of significant value. A far higher, but undetermined percentage have value as fuel for industrial and electric power combustion. In 1968, more than 300,000 tons of aluminum ore and some 25,000 tons of tin were used in cans that were thrown away. Hollis Dole, Assistant Secretary of the Interior for Mineral Resources, exclaimed in a speech that "The resource potential of such solid wastes is so obvious that you are forced to wonder why we call them wastes. There they lie, tantalizingly available, needing only a few technological refinements or perhaps a slight restructuring of the scrap industry to bring them back into the manufacturing cycle." During solid waste hearings, Dole was more explicit. "Disguised as waste," he said, "the ordinary household and commercial refuse collected annually contains nearly ten million tons of iron and steel scrap, almost one million tons of nonferrous metals, including such valuable materials as aluminum, copper, lead, zinc, and tin. In addition, fifteen million tons of glass and lesser amounts of other worthwhile materials are recoverable from these two types of municipal wastes alone, and the energy value of the waste is equivalent to the heating value of sixty million tons of coal. The total scrap and energy value that could be derived annually from these wastes approaches one billion dollars. The potential scrap value of the iron and steel alone is nearly two hundred million. In addition to the recoverable metal and mineral value, each ton of refuse represents a potential of ten million BTU's of energy."

WHAT IS BEING DONE?

One's national pride suffers a bit to hear that European cities are making electricity and heat from refuse and that the horticulturally minded Dutch have compost plants producing soil conditioners out of the organic content in garbage. Why aren't we doing this?

The answer is that while we have the broad management knowledge and technical capability, we have not enacted programs and laws providing incentives or penalties to prod necessary action. The federal program, as described previously, is nothing more than a research and development commitment. Local and state governments have failed to plan on a regional, area-wide basis to take advantage of waste solutions that are only economical on a large scale.

The Bureau of Solid Waste Management has little money to spend and has been a virtually forgotten wing of HEW where the issues of civil rights, education, and health have occupied all of the attention in recent years. (In 1970, the bureau's expenditures were $15.3 million out of $19.8 million originally authorized by Congress. In five years, expenditures have totalled 60.3 million against 80 million authorized.) Under the 1965 act, both the Federal Water Pollution Control Administration and the Bureau of Mines have been empowered to perform waste disposal studies and authorize research programs, but these activities also suffer from lack of emphasis and support. Back in 1966, Athelstan Spilhaus urged in a National Academy of Sciences Study (*Waste Management and Control*) that a model city incorporate a prototype waste disposal system, using all the latest innovations, but this has never even been begun. The Department of Housing and Urban Development does provide limited funds through "701" grants under the Housing Act for community solid waste disposal systems, but little progress is reported in this sector either.

Richard D. Vaughan, the patient director of the HEW program, notes in the preface of a study on European waste practices that these countries "are not using an advanced technology. Instead," he adds (with tongue-in-cheek?), "the economic situation is such that a more expensive system for refuse disposal can be justified." This means that nations with a lower level of technology and no higher a standard of living—at least in conveniences and luxuries—attach a higher priority to their natural resources and environmental cleanliness than we do.

Better Dumps

There are quite a number of stop-gap solutions that do not require any new machinery or twentieth century wizardry. One that is being tried was conceived to relieve the shortage of land available for dumping sites. It amounts to nothing more than

making a mountain of "sanitary landfill" and eventually planting vegetation on top of it to make a scenic area or park. In
flat country, "garbage mountains" could provide landscape diversity and hills for winter recreation. Such projects are being
undertaken in quite a few cities. A ski hill is planned in
Wheaton, Ill., near Chicago, and in Virginia Beach, Va., refuse
from the Norfolk area is being used to make an amphitheater
and roller skating complex on top of a peak of trash sixty feet
high and one thousand feet in diameter. The hole from which
dirt is scooped to keep the trash covered is being made into a
lake. The banks of the hill have been landscaped to prevent
erosion.

With no more room for dumping in its immediate environs,
Philadelphia has begun to haul waste via the Reading Railroad
to reclaim ruined countryside gouged by strip mines. The so-
called Garbage Express costs the city about 25 percent less per
ton of waste than incineration. A similar project, backed by
HEW, is being tried at an abandoned strip mine in Allegheny
County, Md. If cities can work out the transportation logistics
(no small feat), there is ample opportunity for land reclamation
through refuse throughout the Appalachian country and the
hillsides of West Virginia, western Maryland, and Pennsylvania, as well as in other scarred regions of the nation.

Recycling

One of the obstacles to reclaiming wastes is the general
practice of mixing them all together. Thus it is only industries
with a homogeneous batch of leftovers who find it practical to
recycle. The answer would seem to lie in dumping household
wastes into separate containers—for paper, glass, tin and aluminum cans, vegetable and other food scraps that could be used
in compost—except that this is an inconvenience and the costs

of collection go up. Even so, some communities (e.g., Madison, Wis. and San Francisco) have encouraged separating wastes such as newspapers and cans, after ensuring that there was a market for these materials. The drawback is that these efforts have been costly and have not directly rewarded the homeowner, so cooperation has not been enthusiastic or widespread. Where large-scale composting of putrescible refuse is possible, there is too often no market for the fertilizer or rail rates are too high. The Eliassen waste study noted that fourteen of fifteen compost plants built in the U.S. during the last two decades have failed. It is possible that someday in centralized areas like cities, apartment houses will contain separate pipelines or containers for food, metal, glass, and paper wastes that will be integrated with the collection and processing system. In rural areas, homeowners may be restricted in the type and weight of their wastes so that they will be forced to stack newspapers in the heavy steel compressor bins that were available during World War II, and dump cans and bottles in other receptacles. The economic implications and possibilities of these and related ideas have not been analyzed at all.

While it would be ideal to separate and channel wastes at the source, it is not yet really feasible except, as noted, in an industry where the effluent is homogenous and concentrated. However, there are several possibilities for sorting our wastes once they have been collected. The Stanford Research Institute has developed what is called an air classification system to separate dry waste particles once they have been ground or shredded. The particles are distributed by weight as they are struck by a column of air under pressure. At the University of Missouri at Rolla, bottles have been crushed to produce an effective road surfacing material.

Vaughan and his engineers pin their highest hopes on two experiments in processing a heavy mixed load of municipal waste. He has waxed most enthusiastically over a truly revolutionary furnace made by the Combustion Power Co. of Palo Alto, Calif. It is free of air pollution and could be on the market by 1974. It consists of an incinerator on a fluidized bed that burns refuse at high pressure, powering a turbine through hot gas discharges and thus driving an electrical generator. The contractor has assured HEW that the furnace could provide up to 15 percent of a community's electrical power or enough heat to desalinate water. Five units would handle all of San Francisco's wastes, forty would take care of New York, at net operating costs of one dollar per ton, compared to present rates of $7 to $10, or more, a ton for incinerators with pollution controls.

The other experiment that HEW has been touting is copied from Europeans. It is a Gondard hammer-type reduction mill, being tried out in Madison, Wis., that processes wastes finely enough to reduce considerably the amount of room they take up in landfill. The idea is to extend the lives of city dumps with such a mill. In over two years of testing, the operation has cost $6.60 per ton of waste and has produced landfill that is comparatively free of both litter and odor.

Autos and Railroad Cars

Disposal of both automobiles and railroad cars presents major difficulties. No one really knows how many cars are abandoned or scrapped each year. At the Senate hearings, Secretary Finch noted that nine million cars are now produced annually, each having an average life of seven years. Some 20 percent are abandoned; 70 percent land in junkyards and 6 percent are salvaged. What is unknown by the experts is how many of the junkyard cars are eventually sold for scrap after being stripped of still useful (or nonsalvageable) parts. For quite some time, auto-shredding plants have been economical, provided they are located in urban centers where there is enough of a junk back-log to keep them operating steadily. The old technique of crushing and baling cars is uneconomical because it does not get rid of all the materials such as glass, plastics, upholstery, and lining, that cannot be taken in the scrap metal process. Congress is considering new legislation that would require car purchasers in effect to put down a deposit redeemable only when the car was turned in to an authorized dealer or a junk re-processer. This or some kind of tax to pay for reclaiming the car is going to be necessary. In 1968, 43,000 cars were aban-doned on the streets of New York, 24,500 in Chicago, and half a million more in other cities, out in fields and swamps, and over handy embankments where they either cost the taxpayer dearly to clean up or became rusting eyesores.

The annual burning of wood from some eighty thousand scrapped railroad cars is a source of air pollution. Incredible schemes have been proposed to prepare these for salvage. None has been tried yet. It has been suggested that high-speed water jets be used to cut the three to four and a half tons of wood off of each car or that emission control stacks be installed and the steel-encased wood be ignited as in a sealed furnace. In either case, afterwards the cars would be ready for scrap steel processing.

Mining Waste

Bureau of Mines scientists appear no less busy in the alchemy of making use of waste products, although one must ask whether their ebullience might not be somewhat strained. After all, the Bureau's main role is to champion the mining industry, whose health is *not* necessarily enhanced by the rivalry of salvage and recovery operations. In addition to getting some funds under the 1965 Solid Waste Disposal Act, the Bureau of Mines is studying reuse methods under the 1969 Mining and Minerals Policy Act.

As a matter of fact, in 1968 BuMines put out a fascinating twenty-eight-page pamphlet entitled *Wealth Out of Waste* that pointed out all kinds of possibilities. Metallurgical techniques are being tried to recover gold dust from the more than half million tons of fly ash produced by municipal incinerators. It is believed gold from this source could supply 10 percent or more of the nation's industrial needs. Assays show that up to 75 percent of all incinerator residues invariably consists of metal and glass. Now researchers are trying to develop separation techniques to concentrate on this stage of disposal and it is reported that "results so far are encouraging." Power plants pour twenty-five million tons of fly ash a year into the air. Yet tests have been successful using recovered fly ash to fight underground fires in abandoned coal mines. Early this year, BuMines announced that laboratory tests had indicated as much as two pounds of fly ash could be applied to a car tire surface to improve traction and reduce skidding without weakening the tire. Fly ash is also being tested as a sidewalk surface.

At present, the BuMines is just completing a pilot processing plant to process one thousand pounds of glass and metals an hour. Other projects underway include recovery of ferrous metals from city garbage; production of synthetic gas in underground refuse, use of air to fluidize and thus separate nonmetallic from metallic but nonmagnetic car scrap, recovery of copper from junk car armatures as well as cast iron from the engines, and the feeding of tires into a heat reactor to recover 140 gallons of liquid oils and 1,500 cubic feet of heating-value natural gas per ton of tires.

Already in the U.S., the salvage industry does between $5 and $7 billion in business a year. It produces more than half the nation's copper and two-thirds of our lead and has limitless potential in helping to reduce importation of other essential metals such as aluminum (up to 80 percent imported) and iron ore (40 percent).

In the summer of 1970 the most severe shortage of scrap in over a decade beset the nation's steel companies. But it was an opportunity not met by salvage technology. This condition cannot go on forever. The Eliassen report concluded that the economy will suffer increasingly from shortages in essential metals if recycling techniques do not become a routine and important element of natural resource production. And referring to municipal wastes alone, the National Academy of Sciences stated that "The two hundred million tons per year of solid waste material represents a national resource and will in time be a major one."

WHAT MUST BE DONE

Congress is presently considering the Resources Recovery Act of 1969, sponsored by Senator Muskie, and the title indicates that the emphasis in the future solid waste legislation is to be on recycling and reclamation. The proposal also includes an ammendment by Delaware Sen. J. Caleb Boggs to create a national commission on Materials Policy that would report by 1973 on all of the alternatives regarding recycling, and reuse of waste. If not watered down by a House measure, this bill would triple funds for federal efforts against waste. As this was written, it seemed likely that a bill would emerge from Congress that expanded significantly the scope of the 1965 Act by providing municipal planning grants that could be followed up by demonstration grants. In other words, a city or regional government would be given money to come up with a "system" type approach to solid waste, not merely a proposal to treat refuse in a certain ingenious manner but a comprehensive plan encompassing control of the waste load at the source by recycling, streamlined collection of what was not reuseable and a scheme for treating or disposing of these wastes that took into account environmental, land-use factors. Then if this plan satisfied the federal government's criteria and seemed exemplary in all respects the applicant would receive a grant to finance a prototype system. This would bring close to fruition Athelstan Spilhaus' proposal mentioned earlier to build a waste management system from scratch in a model city or new town.

The federal government clearly has the advantage in being the innovator and leading the way in coping with the garbage crisis, but there is still the possibility that as this happens there will not be a tandem effort to reduce waste at the source. While certainly much greater federal participation is needed, it is just as essential that Congress enact laws to tax or in some way penalize industries, institutions, cities, and even homeowners for any waste they produce that is not to be reused. Along with this should come legislation, such as the proposal by Wisconsin Senator Gaylord Nelson, to reduce the incredible amount of paper packaging and other containers.

Nelson has proposed a bill to set national packaging standards and then require industries to pay solid waste fees for products that cannot be recycled or do not degrade easily when disposed. The measure, entitled the Packaging Pollution Control Act of 1970, would empower the federal solid waste agency to set rates and schedules for compliance. Obviously, the penalties would be higher for nondegradable items than for those that decompose quickly.

Paper alone is a major contributor to the garbage crisis, composing in weight 50 percent of the urban collection. Well over fifty million net tons of paper products are used annually and the best evidence is that more than 80 percent is used only once and discarded. Newsprint, which can easily be reprocessed, makes up about ten million tons. The NAS waste study con-

cluded that "There are no major technological limitations to
the reuse of newsprint and paperboard" (taking up to thirty
million tons). Magazines using coated paper and special inks
present difficulties, though the obstacles are not insuperable.

Plastic materials, increasingly prevalent in packaging, are
most troublesome. They do not decompose in sanitary landfills.
Worse, says Richard Vaughan, "The burning of polyvinyl
chloride produces hydrogen chloride; this has a corrosive effect
on the firewalls of incinerator units and can necessitate expensive repairs." Obviously the plastic industry will have to
develop packaging that is biodegradable and that does not emit
harmful gases when incinerated. Better yet would be legislation to reduce the excess of packaging that presently is an
unnecessary cocoon to consumer products and, for example,
comprises 40 percent of the cost of a can of beer. "I wonder
whether the purchaser knows this cost," Vaughan asked a conference gathering, "and if he knows, does it influence his
purchasing habits?"

In Chapter 1, I endorsed the proposal by Senator Proxmire
of an effluent charge—at least the principle is correct, although
it needs refinement and further debate as to its economic effects.
An effluent charge will simply have to be levied sooner or later
on all industrial-consumer products that end up as environmental pollutants. A consulting engineer named Leonard Weg-
man, who is a member of the New York Board of Trade's
Environment Council, has suggested a flat one cent per pound
levy on all consumer products that become waste within ten
years. This would result in a $35 tax on a car, $5 on the
Sunday edition of *The New York Times,* and one-sixteenth
of a cent for a cereal box, and it seems too arbitrary to be
effective. The approach is right, however, and it is undoubtedly
the only way we will ever curb our wasteful ways. Without
such a charge, government waste disposal efforts will always
trail far behind the mounting load. Or put differently, legis-

lation to support recycling grants will not work alone, because it assumes that the volume of waste will increase. This is a dangerous assumption.

Markets Needed

The biggest obstacle to recycling and innovative refuse solutions has been the lack of a market. This would not affect a plant's reuse of materials it previously wasted, but it has deterred companies that are set up exclusively to take solid waste from others and make something of it.

A survey by the *Wall Street Journal* disclosed that companies selling products made from recycled waste were having a difficult time. For example, the Metropolitan Waste Conversion Corp. in Houston was processing 25 percent of the city's garbage to produce paper and other products, but the company was losing money because it couldn't sell its product. According to federal officials, paper companies are not interested in recycling because they are so heavily invested in their own forest supplies and some of the large publishing operations in this country (e.g., *Time, Inc.,* the *Los Angeles Times,* and *Newsweek*) have their own forest products subsidiaries. In short, it has been cheaper for paper companies to rely on raw materials, and thus, according to the *Wall Street Journal* report, perhaps up to 300 million trees were felled in 1969 that could have been spared to meet future shortages had these companies purchased recycled paper. The Houston operation also found no buyers of compost and had success only with sales of metals like copper. Richard Vaughan has complained loudly that the government ought to make freight rates more equitable for salvageable or scrap materials. Present shipping costs for reused paper and metals are, for strange reasons, higher than those for the raw materials. Other experts have said that the government could turn around this pessimistic trend by purchasing for its vast inventories and needs only recycled products if they were available. Both these points—changing freight rates and making a market—are entirely valid. As Vaughan has said often, "Improved technology in the area of separation will have no practical effect unless materials can be sold and utilized."

Some Goals to Set

What goals should be set for the Bureau of Solid Waste Management or, since environmental agencies are presently being lumped together under a reorganization plan, the section of the Environmental Protection Agency that is empowered to deal with the garbage crisis? For one thing, starting at the top, a national masterplan must be devised that will interconnect the problems and remedies of air and water pollution, natural resource supplies, land uses, and ecological considerations with waste disposal. Then some hard-nosed economic planning is needed. It is incredible that through the middle of 1970 there had been no thorough studies pointing out the impact of such measures as an effluent tax, or assessing the feasibility of developing markets to absorb all the intriguing concoctions suggested by BSWM and BuMines. Finally, the Federal waste authority has got to start dealing with regional entities instead of local bodies in order to encourage the only approach to waste disposal that can ever work—the area-wide solution. You read daily about a town rising up in arms because a city or another community has purchased its land for a dump. It is a fact that small towns and cities usually cannot afford the equipment or trained personnel to run a disposal service, and, as single jurisdictions, they often cannot settle on a good dump site.

There are not going to be any magic cures. The Atomic Energy Commission's idea of converting wastes back into their separate and basic elements by use of a "fusion torch" is far down the road, since we have not yet even been able to produce a fusion nuclear reaction. And one should not expect soon that any of the new incinerators or garbage mills will be widely put to use.

In the meantime, you should urge that your local authorities have planned far ahead for waste disposal needs. Ideally, in rural or suburban areas, they should have been coordinating with neighboring towns as to the sharing of incineration and other equipment or in the acquisition of land for sanitary fill. Ask members of your county planning commission, councilmen, and air, water, and sanitation officials whether it would be feasible—or just possible, in hopes of starting something—to establish separate collection of paper, cans, and glass (and putrescible matter, if you are in an area where compost can be marketed). These questions may be more naturally posed outside of the city where conditions in the sanitation system are less chaotic and there is more time to make adjustments. In most cities, officials are already up to their necks pondering over these questions. However, they are more than anxious to know that if they have to create new inconveniences by forcing you to separate cans from newspapers you will cooperate. You can also be effective telling them and your Representatives in Congress that you favor laws that would reduce packaging bulk and bring an end to no-deposit, nonreturnable bottles and cans as was done in the summer of 1970 by the city of Bowie, Md. However, outright banning of such containers will not solve the problem because a lot of people in this affluent society will still throw away deposit bottles. The best solution would be a deposit charge high enough to keep the bottles and cans cycling, a fee maybe as high as ten cents.

There are several new gadgets on the market to "compact" or reduce the volume of trash in the home. These new appliances raise some questions that are best answered by your sanitation officials. For example, while they may save the trash collector some difficulties in handling bulk, they present problems in treating the wastes later, since everything is squashed into a plastic bag. These compacters are being touted for apartments and city residences on grounds not completely environmental but for reasons that are equally persuasive. They save you trips out the door where you might get mugged.

A year ago Reynolds Aluminum encouraged Boy Scouts and civic groups to collect cans by offering a half cent a can or $200 for a ton (forty thousand cans). Adolph Coors Brewing followed suit. Coca Cola has reintroduced a pint-sized (5¢) deposit bottle. Perhaps these or other companies would be willing to cooperate with your town to encourage the mass collection of cans or bottles.

Like it or not, we—all of us—are the problem and the solution to the waste tragedy. As stated by the Eliassen report, "Clearly, comprehensive waste management is a legitimate and necessary expense of living, a fact not yet appreciated by the public."

6 ❧ WASTE LESS

A Checklist for Individuals

MOTHERS HAVE OFTEN resorted to that old homily, "Eat all your vegetables, dear. Think of all the starving children." The recalcitrant child might or might not eat all those vegetables, but he very soon would realize that any leftovers from his supper were unlikely to find their way to starving children anywhere else in the world.

One certainly wouldn't applaud the implication of the desperate mother's plea that it is only by not taking complete advantage of their abundance that those with plenty are being unfair to the have-nots. But the main point—the imprecation against wastefulness—is at the heart of the matter.

We live in a world increasingly conditioned by our wasteful habits and profligate life styles. It is one thing to criticize *the big polluters* and the seemingly inert politicians, and to bemoan the direction that growth and consumption have taken in this country. It is quite another thing to examine one's own personal contributions, to confess like Pogo that "We have met the enemy and he is us," and decide precisely here and now what individual sacrifices are to be made.

As it is put so directly by Con Ed President Charles Luce:

We order our priorities every day. We order them when we decide whether we will ride on the subway or in our automobile. We order them when we decide to buy a Fiat or a Volkswagen or an Oldsmobile or a Buick. We order them when we decide to install an air conditioner in our home. We order them when we decide whether we are going to wash our laundry with soap or detergent. We order them when we decide whether we are going to buy beer in noncorroding aluminum cans that are detrimental to the environment. We even order our priorities when we throw waste paper and cans and empty cigarette packages in the street. . . . What hope is there for reordering our priorities if we are not just a little bit tidy in our individual lives?

IF EACH PERSON . . .

The responsibility of the individual in the environmental crisis has been recognized in widely differing quarters. "I know it may sound awkward, but we've got to develop environmental quality consciousness in every citizen," Maine Senator Edmund Muskie said to me one day just as President Nixon's Environmental Quality Council had come to realize that its most difficult task was to inculcate this sense of responsibility in every citizen. There is already a growing movement among young people to opt out of the established system to seek a simpler, more frugal, often bucolic existence. Many conservation organizations and small citizen-action groups put out check lists of things you could do throughout the day, at home, commuting, and on the job. What follows are the most practical and worthwhile suggestions from these, plus some of my own. This catalogue is by no means meant to be complete. It begs your ingenuity and attention to further possibilities.

Some proposals may seem ridiculous. At least one newspaper reviewer has made fun of the Sierra Club's suggestion in its book *Ecotactics* that people lay bricks in their toilet waterclosets to save water when flushing. But this suggestion has substance. American toilets were designed with no thought given to waste efficiency. As a result, they use several quarts more water than necessary to clean the bowl. Theodore L. Mizaga, a member of the Prince George, Maryland, county planning board brought this to the attention of officials dealing with the water shortage around metropolitan Washington. He estimated that if the bricks saved two quarts and each person averaged seven flushes a day, the saving would be three and a half gallons or over ten million gallons daily for the more than three million Washington area residents. This would reduce the strain on small sewage treatment plants, argued Mizaga, and would save a family about five thousand gallons a year. As reported in the *Washington Post,* Mizaga and his planning board associates hoped they could get plumbing manufacturers to design a more efficient toilet.

The operative phrase in Mizaga's calculation is "If each person. . . ." The problems of waste management, pollution, land use, and so on, are so vast and complicated that altering one's individual habits may seem insignificant—even silly— when suggested as a solution. Why should you put bricks in your toilet when the man next door freely uses a hose, rather than a broom, to clean his driveway? The best answer, as suggested in another connection, is: If you are not part of the solution, you are part of the problem. If you and I don't make a personal beginning, who will?

SAVING WATER

❈ So try putting the bricks in each watercloset. It is possible to accomplish the same result by bending the water tank float rod downwards to make the toilet valve close sooner, but if you bend it too far the toilet will not flush completely. (And so, you cynics who expect weightier wisdom from the Sierra Club, back to the water closet.)

❈ Turn off the faucet when you are brushing your teeth.

❈ Do not use the dishwasher or the laundry machine until

they are full. Rinse all the dishes at once before filling the dishwasher.

⚓ Do your utmost to conserve water during spring and summer warm weather. In Fairfax County, where I live, the Water Authority has urged that lawns be watered "only when needed and before 4 P.M or after 10 P.M.," even-numbered addresses on even-numbered dates and odd-numbered on odd. The authority has also advised keeping cold drinking water in a bottle in the refrigerator rather than running water until it is cool enough, and taking showers (using about ten gallons) instead of baths (thirty-six gallons.)

⚓ In hot, dry climates, consider the advantages of planting less lawn and more shrubs and plants that need less water. At the same time, you will produce scenery that is native and in keeping with your natural environment.

SAVING ELECTRICITY

The energy crisis is severe, particularly in summer when air conditioners strain even peak load capacities. It is quite possible that utilities eventually will have to charge customers more, rather than less, per kilowatt hour for increasingly higher loads used. Since a utility saves money by generating power in volume, its excess profit due to the reverse rate scale would have to be spent for the public's benefit or put into some kind of environment trust fund. Until this happens, there are savings possible within the system.

⚓ Do you really need all those high wattage light bulbs, unless you are reading by them?

⚓ Do you really need an electric can-opener or all the other household gadgets, ranging from electric toothbrushes to swizzle sticks and "hot combs" for men to make their hair wavy?

❦ Turn the lights off when you are not in the room.

❦ Do you need to set the air conditioner at such a low temperature, let alone have an air conditioner?

❦ Think of the ways you can save electricity in the office (but don't upset your secretary by taking away her electric typewriter).

The result of saving on power can be multiplied into vast environmental improvements such as preventing the need for a new nuclear plant on your waterway and cutting down air emissions from conventional generating plant smokestacks. The doubling time for population is now about thirty-five years, yet energy requirements double by the decade. Something has got to give.

PREVENTING AIR AND WATER POLLUTION

❦ Consult the list of detergents provided by the FWQA and cited on page 9. Admittedly, it doesn't give you much choice in selecting a cleanser low in phosphates that will not contribute heavily to the over-enrichment of waterways.

❦ You can also try the recommendation of a Canadian women's group, STOP (for Society to Overcome Pollution), and use a combination of soap powder and soda in the washing machine. First you must "strip" your clothes of detergent residues so they will not turn yellow under the new treatment. Do this by rinsing them in hot water with about four tablespoons of soda. Thereafter, use soap flakes or liquid soap and add from two to four tablespoons of baking soda per load.

❦ While I am loathe to tell you to stop using your fireplace to attain a cheery glow on cold winter nights, you should avoid open burning of refuse or garden and wood trimmings. Air pollution officials insist that the smoke and particulates from such fires *do* add up to a significant hazard, even though automobiles get away with contributing a far higher tonnage of air pollutants. Most localities provide collection and disposal for burnable refuse. It can go into a landfill or a clean and efficient incinerator.

❊ In many instances, you are not using enough (if any) of your garden trimmings for compost. This practice will save fertilizer and produce a better organic additive to your garden anyway.

REDUCING WASTE

❊ Cut down on your consumption of paper by sharing newspaper and magazine subscriptions, opening large Manila envelopes carefully enough to reuse them, and saving wrapping paper. Urge local officials to institute public bins for storing newspapers to be picked up by a wastepaper company.

❊ Discourage "junk" mail (which gets thrown out) by marking the envelopes of material you know you don't want to open, "refused" or "return to sender." If you feel strongly, open the mail and send it back, "postage due," with note telling the sender you feel strongly about waste and asking that he take your name off his list.

❊ If possible, buy food in bulk, thus minimizing the amount of packaging.

❊ Reuse plastic bags.

❊ Do you really need all the "convenient" paper products that are flooding the market, from paper plates to sleeping bags and towels? Use rags instead of paper towels for cleaning, for example.

❊ Buy dairy products in glass bottles, not coated wax cartons.

❊ You will find that your supermarket carries few or no

returnable bottles. Tell the manager you wish he'd pass along your views to the bottlers and order "deposit" bottles *if available*. Complain about no-return, no-deposit bottles to the National Soft Drinks Association, 1128 16th St., Washington, D.C., 20036. Find out from the Glass Institute, 311 Madison Avenue, New York, N.Y., 10017, if there is a local or regional glass reprocessing unit in your area.

※ Check through local officials (sanitation or public health authorities) to see what scraps might be collected for recycling by metal and paper reprocessors in your area. Spread the word among your neighbors as to the availability of these services and the possibility of getting assistance from the local authorities.

※ You can make headway in avoiding "hidden" costs by supporting or helping to organize antilitter campaigns. Roadside cleanup alone is known to cost over half a billion dollars nationwide and is rising 10 percent or more annually. According to a report by the National Academy of Sciences, a typical mile of highway gathers in one month 1,304 pieces of litter, of which 59 percent is paper, 16 percent cans, 6 percent bottles and jars, 6 percent plastics, and 13 percent miscellaneous odds and ends. During 1970 state and city officials have been impressed by the results of citizen drives to clean up this mess.

※ Use a shopping bag or basket. Let the market and stores keep their paper bags.

※ Use the backs of envelopes for scratch paper and write on both sides of pads and stationery.

CARS AND BICYCLES

The slower you travel, the more pollutants your car emits, particularly when you are stopping and starting in congested traffic. Big cars are really unnecessary in the city because they never get to use their power. But they are three or more times as dirty as, for example, a Volkswagen. Eventually smaller, specialized cars may be required and downtown or city streets will be closed to auto traffic. Cars basically are a most inefficient way of moving people and are outrageously expensive per mile when you weigh the benefits to the number of people riding in them (usually one or two per car) compared to the costs of the car itself, fuel, then servicing, repairs, and depreciation. It never fails to amaze me to see so many people commuting all alone to a metropolis like New York where cars are credited with over 70 percent of the air pollution. The message is basic.

꙼ Cut down on the effects of your car by selecting one with no more power than you need and then use it as efficiently as you can.

꙼ If you commute by freeway and feel a larger car is thus necessary, try to form a car pool and thus fill the vehicle. Then provide your wife with no larger a car than she needs for shopping. If you live in or on the outskirts of the city, you probably need no more than a small car yourself.

꙼ Must you have two cars at all?

꙼ Have you thought of bicycling, which, as a method of transportation as well as a sport, appears to be enjoying a comeback? Texas Congressman and conservationist Robert Eckhardt bicycles to work on Capitol Hill, finding that it is a clean and invigorating alternative to being couped up in one of Detroit's boxes, lurching and braking in the traffic. If Eckhardt can keep his head clear from the pall of carbon monoxide that hangs over city thoroughfares and avoid being hit by a car, he'll come out ahead of the game. In fact, races held in Washington have proved most conclusively that the bicycle is a faster commuter except on main arteries. It is of course wonderful exercise, recommended particularly by heart specialists. According to the *Wall Street Journal,* stores and banks are now installing conveniences for cyclists (parking stands and special

windows) and in my neighborhood, high school students have
been lobbying for bicycle paths to run beside commuter routes
that daily become more clogged with cars. Remember when
William Buckley was derided for running for Mayor on a plank
that included opening bicycle paths throughout New York?
Well, if that proposal at least had been taken seriously, the city
would be a bit more fun and maybe a mite healthier. Bring
back the bike. It has great merits, as Europeans and Asians
have long known.

※ Take advantage of public transportation. There is no
reason why buses, trains, and transit systems cannot be made
more efficient again if the car were only dethroned as king of
every city approach and thoroughfare. So support bond issues
to finance rapid transit and urge your public officials to promote
more efficient transportation systems.

※ If you are tied to a car and cannot work out a car pool
or some kind of saving with your family or friends, then at
least make sure your machine is as clean as it can be. If it is
a new model with emission controls, check to see that they are
working well. If it is an older model, contact your regional air
quality control board or write to the National Air Pollution
Control Administration and ask what controls you can install.
I cannot vouch for the various antipollution kits now advertised
by auto makers and there may be changes by the time you read
this, so it is wiser to ask the experts.

※ You can also use a fuel without lead additives. As
noted in Chapter 2, no-lead fuels do not corrode exhaust control devices and on the basis of available evidence appear to
produce less hydrocarbons.

※ Aside from keeping your car well tuned so it will not
emit unburned fuel particles, you may be spending money need-
lessly on high octane gas. Oil companies and their service
outlets have been most uncooperative in providing motorists
with octane ratings, but if you can find out the octane content,

you ought to use as low a rating fuel as will not produce
"engine knock" resistance. Technically speaking, the octane
rating assesses the ability of the fuel to withstand heat and
pressure before it combusts. The higher your engine's com-
pression ratio, the higher the octane rating has to be. If you use
too low an octane fuel, your bearings, pistons, and valves will
suffer, but too high an octane increases the emissions of unburned lead additives. Sun Oil Corporation, which offers eight
grades of fuel (at Sunoco stations), recommends that motorists
use the lowest grade that will not produce a knocking or pinging
sound during acceleration. The Federal Trade Commission has
met resistance in trying to require *all* gas stations to post octane
ratings. Your letter to the FTC calling for such a measure
would help support pending legislation.

The federal government and other public agencies have experimented with far cleaner propane burning cars and you can
obtain conversion to natural gas for about $350, but the range
of these cars is limited, as is the supply of fuel, and, as already
noted, this approach is really only feasible for a transportation
fleet (e.g., taxis or government vehicles). In the meantime, use
the smallest engine you need, buy no-lead fuel (having the
engine adjusted to take it, if necessary), and drive and idle
your engine as little as necessary.

HABITS IN GENERAL

≭ Plant a garden or tree. It is good to hear that, according to
various reports, indeed more people are cultivating fruit trees,
flower beds, and vegetable gardens in their backyards or fields.
Ohio State University has printed twelve thousand copies of a
book on vegetable raising (*Home Vegetable Gardening in Ohio*

by J. D. Utzinger, W. M. Brooks and E. C. Wittmeyer) and, as reported in the *Wall Street Journal,* a woman in Lincoln, Nebraska, saves 50 percent on canned tomatoes by putting up her own tomatoes from a fifty- by fifty-foot plot. Richard Reinhardt, an imaginative freelance journalist and author, asked in the spring 1970 issue of *Cry California,* what had ever become of Arbor Day when everybody planted a tree somewhere, right on down from the President of the United States to school children. Resurrecting this national occasion was suggested to President Nixon's domestic advisor this year, but other events got in the way. Americans are so much on the move and our sense of time is so compressed, that few of us stop to consider the value of a seedling five, ten, twenty, or more years hence, even if we're no longer residing near enough to see it. Of course home fruit and vegetables are healthier, or at least you know what kind of fertilizers, pesticides, and herbicides these products have been exposed to. One sad development of public planting programs is the lack of diversity of the shrubs, trees, and flowers. Rather than take the pain to plant an indigenous species, park and highway landscapers too often go to mass-production nurseries and order the stock plants for holding soil or growing quickly. As a result, the green strips in cities and along highways throughout the nation have acquired a somewhat dull conformity. Worse, however, is the effect of such practices of wildlife —particularly birds—who live off of and nest in peculiar native vegetation.

※ Boycott furs and other articles taken from species threatened with extinction. What are these creatures? Check with the International Union for The Conservation of Nature, and the Department of Interior's Office of Endangered Species who keep a file on this subject called "The Red Book." New York City has passed a law banning sale of furs from endangered animals and the federal government in 1969 enacted a new Endangered Species Bill that bans importation or interstate shipment of such animals or their specimens.

❦ Wherever there are bounties for killing animals such as the wolf, so-called "varmits" like the coyote, or birds like the crow, or wherever creatures are considered as nuisance predators, you should consult environmental conditions most carefully. Invariably, the animal that is considered as a pest plays an important role in the natural balance of a particular ecosystem and gunning down this creature is apt to begin a chain of undesirable side-effects such as the multiplication of its prey.

❦ If you need mechanized equipment to work inside and around your house, think of the disturbance you might cause when using these gadgets and try to muffle their sound as much as you can by soundproofing your carpentry shop and mowing the lawn in the middle of the day, not just before supper or early in the morning when people in the neighborhood are asleep.

REPORTING OFFENDERS

It is a custom in many European cities to go up and scold people seen littering. If they pay no heed, they are reported to the nearest policeman. As a matter of fact, antilitter, antinoise, and antismoke ordinances exist at the local level throughout the U.S., only seldom are they enforced or even known by the constabulary. However, the time has come. Like the Europeans, you should report offenders to the police.

❦ Find out what laws exist that enable you to become a watchdog.

❦ Check out provisions concerning land use, zoning, soil erosion, and air, water, and woodland management in whatever literature your county or town has made available. If the laws are not explained or simplified, urge public officials to do so, or contact your League of Women Voters representative and ask her to intervene.

Interior Secretary Walter Hickel has exhorted citizen advocates to take up the cudgel against local polluters by notifying state and federal agencies and writing letters to representatives and officials. "Walk along your waterway and look for obvious examples of pollution. We need citizen sentries," he said in a statement issued during Earth Week. "If people would adopt their favorite body of water, like they would a child, and watch over it, we would have the greatest enforcement army in the world."

Charles Little of the Open Space Institute has suggested that protecting the environment become a people's crusade, rousing the national spirit like the Second World War (when "everybody did his bit, from victory gardens to meatless Tuesdays.") and supported by *environment instead of defense* stamps. Certainly, the stakes are as high—an environment in which man can survive, spiritually as well as physically. And unlike a war, this citizen's effort—from the ground up—should not cost lives.

Throughout this book, there are suggestions for pollution detectives. But it cannot be emphasized enough that government action, at all levels, depends on citizen outcry. Get to know your neighborhood, its ecology, its wildlife and natural characteristics, and when this fabric is in danger of being torn by unnecessary change and wasteful practices, do not hesitate to ask questions, write letters, and sound the alarm. *You* can and *should* exert pressure. It is, after all, *your* environment.

7 ❧ THE LAND

EVERY CHAPTER in this book is, in essence, about abuses of the land. Widespread erosion and runoff are major contributors to water pollution. The siting of industries and power plants affects the air. Proper location of highways and airports reduces noise. Protection of natural settings from the bulldozer has been a preoccupation of environmental lawyers. Waste disposal strains the capacities of the land in many ways. And of course it is our spectacular growth—economic and demographic —that demands more and more room in the first place. So taken as a whole, this book from beginning to end is an appeal to a new sense of responsibility toward the land. Yet, in particular, this chapter suggests ways of implementing what Aldo Leopold defined more than twenty years ago as a "land ethic." His message, in *A Sand County Almanac,* is evoked frequently in the speeches and writings of conservationists in 1970.

Leopold sought to extend the social, philosophic ethic to man's attitude toward the land, an attitude he found "still strictly economic, entailing privileges but not obligations. We must arrive at the understanding," wrote Leopold, "that the land ethic simply enlarges the boundaries of the community to include soils, waters, plants, and animals, or collectively: the land." Americans' feeling for the land was rooted in overcoming the forces of nature on the frontier, but the time has come, said Leopold, when we should develop a new respect for our potential in altering or maintaining the balance of nature. Moreover, we should not count on the government to assume the burden of stewardship that we were unwilling to bear. "Industrial landowners and users, especially lumbermen and stockmen," he wrote, "are inclined to wail long and loudly about the extension of government ownership and regulation to land, but (with notable exceptions) they show little disposition to develop the only visible alternative: the voluntary practice of conservation on their own lands."

In sum, to Leopold, "Land, then, is not merely soil; it is a fountain of energy flowing through a circuit of soils, plants, and animals. Food chains are the living channels which conduct energy upward; death and decay return it to the soil. The circuit is not closed; some energy is dissipated in decay, some is added by absorption from the air, some is stored in soils, peats, and long-lived forests; but it is a sustained circuit, like a slowly augmented revolving fund of life."

Nobody, in my judgment, has phrased the importance and meaning of the land any better than Aldo Leopold. Increasingly, we are finding out that ecological considerations are critical even in the cities and slums. In fact, planners have not yet found an exception to this test of Leopold:

A thing is right when it tends to preserve the integrity, stability, and beauty of the biotic community. It is wrong when it tends otherwise.

While this is so true and perhaps even obvious, the fact remains that the rule seldom has been applied to land use in this nation, and the consequences are all too apparent. We have set aside great national parks and wilderness areas as the places where the so-called "laws of nature" will be observed—sacred jurisdictions beyond the reach of our rapaciousness. But simply having these glorious sanctuaries has helped to lull us into thinking we could do what we wanted with the rest of the land without making sacrifices. And now as we flood the parks with automobiles and send our vacation brigades into the wilderness, we ought to realize before it is too late that only by treating the entire country as a park are we going to cultivate the land ethic.

Thus the main question of this chapter is how do we go about ordering development—the use and management of land

—without toppling the life pyramid or cracking it, so that enjoyment of the land endures. In seeking an answer to that question, this chapter divides the land into three regions where everyone—unless he is lucky enough to be a park ranger— lives and works: the cities, the suburbs, and the rural spaces. These regions also provide distinctly different ways of life. But of course their economic and social characteristics are intertwined.

The pattern of development in this country has really gone full cycle. As soon as they can afford it, many urbanites flee to the suburbs, eroding the city's tax base by abandoning it to the poor. But, as suburbanites, they continue to rely on the city as a place of work and entertainment. Feeling claustrophobic or seeking variety in their monotonous, unordered new surroundings, these suburbanites are apt to buy a "second home" for recreation or retreat further out in the country, causing a flurry of land speculation wherever there is pleasant scenery, boating, fishing, hiking, or sailing. Rural inhabitants, unable to scratch a living from the land because of rising and disproportionate property taxes or new and specialized technologies, migrate to the cities.

Zoning laws and public land policies must be changed to deal with these and other equally disruptive patterns, and this chapter will touch on both the role of zoning and land laws in attaining that new land ethic.

THE CITIES: PROBLEMS

At the present projected rate of growth, the United States could accommodate a new city the size of Denver (512,000) every two months between now and the year 2000. In 1970 about 140 million of the 206 million Americans lived in or right

around cities (i.e., in metropolitan regions). By 1980, urban expansion will take in another forty-five million people and put
down at least one million acres in new pavement and buildings.
The Urban Land Institute predicts that by 2000, 90 percent of
the U.S. population will live in metropolitan regions (v. 70 percent now) and 60 percent on only 7 percent of the nation's
land, concentrated in three megalopoli—a triangle joining
Chicago, Maine, and Norfolk; from north of San Francisco
south to the Mexican border; and all of Florida. This eventuality
simply must be thwarted by restraints on growth and by
incentives to population redistribution. But you can't just come
to a full stop in an instant, or move people around like toy
soldiers, and so any discussion about the future of cities must
open with gloomy forecasts.

The prospects for acceptable living and working conditions
in the nation's metropolitan regions are frankly not good. The
urban environment already is corroded more than any other and
cannot help but get worse as more people are jammed in. Automobiles cause congestion and air pollution; industrial, residential, and utility plant furnaces, as well as incinerators, emit
more noxious gases; the din on all sides and overhead is
incessant; and the crisis caused by litter, abandoned cars, and
general garbage is overwhelming. Evidence accumulates that
cities are in many ways health hazards. The drinking water,
according to federal health officials, is frequently unsafe or
dangerously overchlorinated. Incidents of food contamination—such as botulina and salmonella—are common. During peak commuter hours in all cities, and throughout the
day in some, the levels of carbon monoxide from cars is well
above that recommended as minimum sustained exposure (see
Chapter 2). Consumer legislation has over the years made the
world safer and healthier. Modern medicine has done the rest
to obtain a sanitary environment. But scientists warn that
crowded living conditions and misuse of land in the cities could

produce more epidemics of disease in coming years. As previously noted, air and water conditions in cities have already undermined public health.

What is generally overlooked is that a city links together a vast and intricate network of environments or ecosystems. To begin with, consider the land used somewhere else (on the average of two and a half acres per person) to provide food for each U.S. city dweller. Then enormous areas of land are required to fuel and provide resources for urban industry. As previously noted, the city consumes mammoth amounts of water and generates mountains of waste daily. The city's fabric of survival and support extends far and wide. As succinctly put by ecologist Raymond Dasmann, "When we start unraveling the web of a modern metropolis, we find that its strands are attached all over the earth."

The increasing unmanageability of cities, the ease and frequency of breakdown in the urban apparatus, from garbage strikes to power failures, is now a familiar theme. Land use trends have been a major cause, since festering residential neighborhoods have been forced to grow up around manufacturing plants, warehouses, shipping facilities, railroad yards, and the smoke-belching electric plants. Rarely have city planners broken this pattern by locating a new civic center, mall, or clinic-hospital complex in a slum except as part of urban renewal schemes which destroyed the old neighborhoods entirely.

One of New York Mayor John Lindsay's more perilous moments came the night the Mets were being toasted as baseball's World Champions. Confetti, ticker tape, all kinds of celebration litter, covered New York like a freak blizzard. Only days before an election, Lindsay would surely suffer if all the garbage were blowing in the next morning's wind. (Besides Lindsay had identified at every opportunity with the underdog Mets.) All night he had to move around the city exhorting district sanitation workers and, working around the clock, they cleaned up New York. Such is the fragility of the modern urban environment that events such as this are becoming routine.

Yet despite these traumas, the city remains the great proving ground of technological invention, scientific research, cultural expression, and free enterprise ingenuity. Free thinking and general nonconformity are encouraged in cities, not stifled as in suburbia. They are, in short, where most people live and where most of the action is. There is a tendency, though, for planners to talk in terms of abandoning the existing cities and building new urban complexes from the ground up, making smooth working machines with all their land used efficiently. Such visions are at least premature and, more likely, narrow and unrealistic. They overlook the metamorphosis of myriad influences that over long periods of time make cities fall into place and generate their vitality. More to the point, these grandiose plans for a million people at a scoop are escapist, in effect, admissions of failure with the cities we've already built. They are typical of the frontier ethic and the misguided feelings about land use that got us into the mess we're in. It simply won't do any more to concede that it is easier to build a new nest than to repair the old one. There is much that can be done to rejuvenate cities—particularly in the way of land reforms.

THE CITIES: PROPOSALS

It is pure myth that cities have no more space within their borders on which to build new housing or locate parks and plazas. Crowded as they are, the big cities can find a surprising amount of room by clearing away derelict buildings, using air space over streets or building on water, obtaining surplus or unused government land, restoring rundown blocks and decaying waterfronts, taking advantage of alleys and small vacant lots, and enacting measures to relieve traffic and parking congestion.

Jane Jacobs, an imaginative champion of the city (see Bibliography) found that there was plenty of land in junkyards, abandoned transit facilities, and a variety of nooks and crannies to solve urban housing shortages. For example, in a survey of only twenty-two out of eighty-seven districts, the New York Metropolitan Council on Housing discovered it could build 100,000 new family units. However, to do so would require drastic overhaul of inflexible, stultifying building codes and money-lending stipulations, as well as regulations permitting new construction techniques and experimental housing.

Presently, during an extreme national housing crisis, many blocks of old housing developments (of twenty years vintage or more) are being abandoned. Why? For one thing federal family assistance plans encourage movement to the suburbs. But basically, these housing rows were shoddily built under tax laws allowing developers to write off rapid depreciation. Now it costs too much to repair or rehabilitate these dwellings, so they become inner city ghost towns. Renewal (to many urban thinkers an anathema when it only displaces the poor) and experimental housing (e.g., factory-built and quick) in these areas would have a good effect by shoring up the tax base and holding together dying neighborhoods.

There is also an as yet unassumed opportunity for industry to initiate or take part in neighborhood restoration by purchasing dilapidated housing and applying ingenuity in replacing it, possibly even integrating residential and industrial complexes so that, for example, the housing can use factory waste heat in return for providing a new job market. The costs of allowing slums and rotting tenements to degrade still further have not been appreciated. Not only is the city deprived of tax revenues as these poor neighborhoods and ghettoes get worse, but rising crime and unemployment adds considerably to city expenses

because these conditions require additional fire and police protection and garbage, social, and welfare services.

The Jungle of Highways and Streets

Highways, streets, and parking lots take up to 70 percent
of the land surface in many cities. Freeways have destroyed
precious vistas, depressed property values along their paths so
that blighted strips sprout up, have ripped open parks or run
over historic landmarks, and are rarely tied into city plans. All
this has been encouraged by provisions of the federal highway
laws that provide state governments with 90 percent of the
funds for interstate roads and 50 percent on other federally
aided roads. But more governors should follow the lead of Massachusetts' Francis Sargent in calling a cessation on interstate
highway construction in Boston until that entire metropolitan
region's transportation needs and patterns are assessed and
plotted far into the future. In New Orleans, San Antonio, San
Francisco, Seattle, and Washington, citizens have staged successful holding actions against freeways by defying state governments anxious to keep federal money flowing, or by citing environmental criteria in recent highway legislation. One hope, as
yet unfulfilled, is that rapid transit systems will be financed and
built and will cure city traffic woes. In Baltimore, plans for a
freeway were scrapped in favor of a broad scheme that will
integrate the road system less disruptively into the city and will
be coordinated with neighborhood and commercial development
and renovation. The secondary development will actually cost
far more than the highway itself.

Consider that up to three hundred square feet of maneuvering
and parking space is required for every car that commutes to the
city and you will agree that curbs on cars must be imposed
soon, even if painful temporary inconveniences follow. City

garages are of course mostly a waste of space and taxwise it usually pays a speculator to sit for a long time on a vacant lot or old property, using it for a parking operation until he can make a killing selling it for high-rise development.

From the standpoint of gaining more land for people to live, stroll, and play on, at least closing off street systems for plazas has got to be done more and more. City planner Lawrence Halprin's conversion of Nicolet Avenue, Minneapolis, into a landscaped pedestrian mall, providing freedom of access and movement that pleased merchants and uplifted people, is a good model. New York Mayor Lindsay has been mulling over ways of coping with cars to give the city more breathing room (and better air). Imposing a differential toll was one thought. Commuters entering New York alone might pay $1.00, whereas the rate would be reduced an increment for each passenger (e.g., would be 25¢ for a four-man car pool). But the idea ran afoul of state control of the toll authority. Another thought is to ban private pleasure cars from 59th street to the tip of Manhattan at least by day. Another suggestion is special city cars—small low-powered vehicles taking far less room than today's monsters from Detroit. But just at what point around the city these cars would begin to operate and their obvious limitations raise many questions. It is proposed here that, at the very least, cities inventory their streets and relate traffic and all transportation to land and health considerations. (Not the least of highways' damage is done by carbon monoxide wafting over the regions through which traffic flows.) Cities must also contemplate more imaginative use of air space over highways, for apartments and commercial and industrial buildings, provided they are well insulated from noise. Pittsburgh and Richmond, Virginia, for example, have used air spaces effectively for plazas and landscaping in old neighborhoods. An 811-unit, cooperative apartment has been begun over the Long Island Railroad tracks in Brooklyn. The main warning that seems warranted in planning new uses on or above streets is that too often such plans are conceived to benefit the city commercially, to lure shoppers back downtown, to attract tourists. The emphasis is on the connection between esthetic and economic vitality, ignoring the social, human values enhanced by giving land back to people. Cosmetic improvement is desirable only if it is not at the expense of urban residents who are not considered cash crop. One way of avoiding this pitfall would be to close off streets or use air spaces first in poor neighborhoods or the ghetto, then place civic and cultural facilities in these projects to make them come alive, hold and attract people. Such rehabilitation should not provide housing that the neighborhood cannot afford. But neither should it be the sort of building that falls into ruin in every city. In the end, everyone will benefit—developer, resident, and city.

It should be obvious that the social and economic benefits from reducing traffic are considerable even if they have not been quantified to a great extent. If a city government had the option of spending that 90 percent federal interstate highway money on whatever transportation system best suited its needs, the city would be way ahead. In fact, to stretch a point, it would be just as good if that money—at least the same sum through a different kind of grant—were spent on improvements that would hold people in the cities instead of encouraging (or driving) them into the commuter fold where they will be spending more on cars or some kind of transportation forever.

Seattle designer-builder Morley Horder has a fine idea for solving still another highway crisis in that city, where surely roads and parking lots already must take up two thirds of the land. Instead of shooting ten more lanes of freeway across Lake Washington into the city (which is now connected to eastern suburbs by two floating bridgeways), why not spend less money building a great mall along Seattle's dead waterfront, extending it on concrete out over the edge of scenic Elliot Bay. There, facing the spectacular Olympic Mountains across an open expanse of salt water, a town would be built within the city— apartments, townhouses, low-cost units, interwoven along the curving bayside with pedestrian and shopping plazas, a maritime museum, restaurants, and other assets. Then, says Horder, smiling over this vision, for he would love to move back to the city, Seattle would glow at night as a place where *people live* as well as work, and the suburbs could keep their ravines, watercourses, and groves of Douglas firs.

Actually this is not as abstract as it sounds and I will leave roads—the major urban land factor—after a few more statistics. These ribbons of concrete that snake everywhere are the biggest plunderers of the city's tax base. They use up hundreds, nay thousands, of acres of valuable land. An ordinary interchange will use up to ten acres. In downtown Atlanta, reports Wolf Von Eckardt, architectural critic and author, an interchange cloverleaf takes up 140 acres of prime land. All this to funnel cars into cities where, according to a national survey, the average auto speed is barely over ten miles an hour. It costs up to sixteen million dollars—sometimes more—to build a mile of city highway. Wouldn't it be less costly to build that mall, giving the city the economic and social benefits from people in perpetuity, instead of the drain from cars that may someday (not far off) be obsolete?

Open Space

A century ago, Frederick Law Olmstead designed Central Park on a seven hundred-acre swamp that had been used by squatters. It was, wrote Olmstead, who afterwards put his visions into more than a dozen urban parks, a place for city folk who couldn't afford the White Mountains or the Adirondacks to benefit from nature, "God's handiwork." A believer in the doctrine of public trust, Olmstead maintained that "the enjoyment of scenery employs the mind without fatigue and yet exercises it; tranquilizes it and yet enlivens it; and thus, through the influence of the mind over the body, gives the

effect of refreshing rest and reinvigoration to the whole system."
Over the recent years, various professional studies have shown
that in dense downtown areas there is a definite correlation
between poor mental health and lack of open space or greenery.
And now ecologists have pointed out that parks and recreation
areas not only are crucial to the human state of mind but are
vital elements of the city ecosystem. This is not a new observation. It is just that we have never taken it seriously. As a result,
life in the cities has become increasingly miserable. There are
not nearly enough Central Parks in America.

Trees and shrubs, patches of field and woods, absorb
noise and traffic odors, screen ugly structures and streets, offer
shade and room to stretch, and please the eye as a break in the
rigid, geometric order of buildings. But they also hold water
and have a "greenhouse effect" by converting great outpourings
of carbon dioxide in the city back into oxygen. Finally, these
open spaces are important refuges for wildlife (mainly ducks
and other birds).

There is unlimited potential in developing much more land
in cities as "open space." Secretary of the Interior Walter J.
Hickel has proposed a great system of urban parks, "bringing
parks to the people," that would tie together large tracts of
government lands that are now unused or outmoded for use.
Following his lead, the administration has called for a review by
all federal agencies to determine what lands, surplus, defense
establishment, Department of Transportation, and other hold-
ings, might be turned over to the public for parks or recreation
centers. President Nixon in his State of the Union Address
noted that one third of the U.S., over 750 million acres, was
federal property and that much of it lay in metropolitan regions
"reserved for only minimal use." But since then the White
House has not given the idea priority. In the summer of 1969,
Hickel had announced plans for the first such urban park, on
either side of the entrance to New York Harbor, taking Sandy
Hook, Rockaway Beach, and possibly later including the city's

Jamaica Bay preserve. This "Gateway National Recreation Area" would ultimately cost about $80 million to tie up and develop —a fraction of its worth if the land were bought out of private ownership. Hickel's total scheme for the cities entails just over five billion dollars in proposed federal grants.

Rather than wait for such programs to work their way through Congress, individuals and citizen groups should begin to inventory public lands in their environs and single out the acreage that could be much more efficiently used and more valuable in a park system. Much of this land is hidden and overlooked and will take a lot of digging to discover. In many cases, you will have to lobby to justify the lands being transferred. The Department of Defense, for example, sits on much of the best city land—in obsolete storage depots, armories, and the like, but the generals will always chorus in effect that "this land is absolutely essential to the national security and will be used in time of war." It would appear, in these thermonuclear times, that a major declared war wouldn't last long enough to need those lands or any other part of the earth, for that matter. And during a major undeclared skirmish in southeast Asia, these defense properties continue unused or obsolete. Often they surround traditional "forts," old creations, strategically overlooking an approach to a city. Modern warfare has bypassed them but their original qualities make them ideal for parks. Boston, New York, San Francisco, and Seattle are cities where splendid opportunities exist on waterfront tracts and islands.

Both states and the federal agencies provide aid to communities and cities that want to acquire open space. Conservation organizations, foundations, and land trusts also play a major role in land acquisition and land use planning. The success of these private action groups will be described later in this chapter. And zoning, tax incentives, regional plans, and land banks will also be treated separately as ways of guiding development or protection of land.

Roof Gardens

There is one ideal base for gardens and greenery in cities that has hardly been tapped at all: the rooftops. Where they have been planted (e.g., New York) they have been a great success. They add an exhilarating new dimension to the city, aside from providing residents opportunities to work with their hands and take pride in growing something. Lawrence Halprin wrote in his 1963 book, *Cities,* that "Up high on the roof there are views over other buildings, sunsets to see, a relaxing freedom from cars and other traffic, a privacy and intimacy which no other city facility can bring, and all of which are difficult to achieve at street level. As our modern cities build higher, and as open spaces become more difficult to acquire, the use of the open space on the roof—which normally is wasteland—becomes more imperative." Halprin has exploited the concept ingeniously in many of his projects, and close to the ground, as in San Francisco's Ghirardelli Square (a roof). Gardens provide a protection from wind and sun and can be structurally incorporated in buildings without much extra cost. Even older buildings, if the traditional building regulations were overhauled, could handle simple gardens.

Federal Grants

Federal grants for urban open space are available under the Department of Housing and Urban Development's Open Space Program. These are 50 percent matching grants stipulating that the applying community must fit its acquisition into a comprehensive plan, that the land be open to *all* the public and that it not later be used for a public facility such as a dump or sewer plant. HUD also extends "701" planning grants under the Federal Housing Act that pay up to two thirds of the cost of a regional masterplan. It is discouraging, though, that HUD does not have its own comprehensive plan for doling out money, but takes action on a first-come, first-served, basis.

Back on the good side is HUD's new "Operation Break-through," a program added to the catalogue of housing assistance plans. Its objectives are grants to developers and housing systems manufacturers that will produce innovative new housing complexes to go up quickly and last and appeal a long, long time. In order to implement this program, however, many state and local building and land use barriers must be broken down. California has already passed a Factory Housing Law that pre-empts local building codes and inspections. Ohio has gone further by creating an office to review and test new housing methods and imaginative land use.

HUD has also broken new ground with an urban renewal demonstration grant to New York to explore ways of using streets, closing them off, building over them or whatever, in order to enhance the environment. This department has of course for some time granted 50 percent money for "beautifica-tion," or landscaping, along city streets.

Through the Land and Water Conservation Fund, adminis-tered by the Bureau of Outdoor Recreation, states can apply for money to purchase parks. The states of Maine, Massachusetts, Rhode Island, Connecticut, New York, New Jersey, Pennsyl-vania, Florida, Michigan, Ohio, Wisconsin, Illinois, Minnesota, California, and Washington, in addition, have their own open space programs.

The most cynical judge of the evidence so far—the success of conservation practices in some cities and the failure to use land efficiently in others—would have to agree with HUD land development official Dwight Rettie, who wrote: "Let us apply the same systems of values to the quality and character of urban life that we apply to the protection of parks and refuges and wilderness and forests and farmlands." Then toss in Lewis

Mumford's observation from *Sticks and Stones* in 1924 that "A city, properly speaking, does not exist by the accretion of houses, but by the association of human beings."

To which we add this. There is plenty of land in the cities, if it is managed with care and forethought. And if it is, human beings will not only have the constant draught and relief of nature, but their associations with one another will become far more stimulating.

THE SUBURBS: PROBLEMS

Ringing every city is a belt ten, twenty miles or more wide—chaotic, undisciplined sprawl—where we have taken out our historic urge to own our very own lot, house, yard and garden, garage, two cars, lawnmower and other fancies, from barbecue to a boat sitting on a trailer. This is not to suggest that having these things is wrong but simply that the lack of order or unity in pursuing and attaining them has produced an ugly sprawl. In these suburbs, our failure to manage land effectively has been the most pronounced. Even if the cities make unprecedented progress toward solving their land problems, the pressures on the suburban fringes of metropolitan regions (which include most suburbs anyway) are going to be terrific. Sooner or later, come what may, great sacrifices and revolutionary changes in land laws are inevitable. Of course they will be less painful and less costly if begun now. It is in the suburbs where the "land ethic" faces its greatest test.

I have tried to show how cities can consume less open space, or have more room for parks, by using their land efficiently. But whereas city governments have some grip on their future, the flung out suburban jurisdictions are a crazy patchwork of micro-governments. Washington Governor Daniel Evans wonders how anything gets done in King County, the largest district in his state (containing Seattle) when it includes over one thousand units of local rule. He calls it an "ill-disguised form of government by legal anarchy." And a study of metropolitan areas has noted that King County is not unusual: Chicago's Cook County has 1,113 localities; Philadelphia, 871; Pittsburgh, 704; and New York, 551.

Major obstacles, then, to land use planning in the suburbs, are built into its very structure. The failure of planning is perhaps most obvious here because, unlike the cities, these outskirts have plenty of natural, undeveloped land remaining to be ruined. It is not a question of restoring or repairing misused environments. Frantic and random growth into these lovely places trickles away unchecked, out of control. Land speculation is one result. Usually, for example, you can get the most for your money by leasing a parcel of land to a gas station or some other ugly, temporary commercial development, while retaining ownership for eventual maximum speculative benefit. In McLean, Virginia, an unincorporated town down the road from my house, there are fourteen service stations within a radius of one mile in what can be called the town center, although this intersection is such an aimless meeting point of gasoline alleys and ghastly strip displays that there is no feeling of real community. McLean merchants are now beginning to feel the pinch of shortsighted development, because McLean's services are so disorganized that nearby shopping malls have taken away most of the business. McLean is becoming a blighted area. So, in their own ways, are many suburban areas.

THE SUBURBS: PROPOSALS

Zoning is the restriction placed on the density and kind of development that can taken place on land. It is *not* a guarantee that land will be used in a certain manner, as is the impression when it is said that an area is zoned commercial or otherwise. Zoning in theory is simply a protection against unwanted development and a defensive action allowing some kinds of construction in a given area. But it is so subject to political alteration and repeal as to be useless as an assurance that land will be kept open or developed in a limited way or density forever. It is a fact of life that when development pressures mount and a locality needs the tax money generated by new construction—commercial or residential—there is nothing to stop a "variance" or exception from being issued. The result is "spot zoning," another term for forced surrender or planning by whimsy for the sake of short-term gain.

The National Commission on Urban Problems, chaired by former Senator Paul H. Douglas, was particularly scathing about zoning practices. After analyzing various land controversies— notably in suburbia—the Commission concluded (see Bibliography) that (1) when developers and local authorities unite, as is usually the case, there is no stopping them, (2) community residents seldom participate in planning decisions and do not understand zoning or other kinds of restrictions placed on land development in their environs, and (3) decisions on zoning follow no general principles or guidelines but are invariably "individualized." In short, zoning has failed altogether as a tool to guide growth because it is so arbitrary and capricious.

Over recent years, "cluster zoning" has gained some acceptance as the most desirable means of fitting enough people on a tract of land to make profit while preserving the largest possible amount of open space. Single family units, townhouse apartments, tall condominiums or cooperatives can be arranged close together in an attractive scheme with terraced plazas, shrubs and shade trees, and landscaped walking areas, while the remainder of the tract is left wooded or open in meadows. This open space may be deeded by the developer as a park or, better, consigned to a homeowner's association to be kept undeveloped forever. The main advantage of the cluster plan is the overall harmony achieved in grouping different land uses. Moreover, cluster communities can be interconnected so as to create a larger fabric into which roads, schools, parks, and shops can be knit most effectively. Experience has shown that clustering the living area is less expensive to a builder, not just because of efficiency and savings in the use of materials, but because he is allowed to install more living units than if he spread them out over a conventional grid pattern. The architectural effect should be livelier or less monotonous than the impression of a thick stand of houses deposited over the landscape, one to a lot.

However, most developers still feel that the public prefers the conventional layout even though it provides less running room for children and less privacy from streets and cars. Many town governments are suspicious of anything that appears to be a savings for the builder, and many towns fear cluster zoning as the opening wedge in the urbanization of their "rural" town. As yet, there seems to be a deeply ingrained public possessiveness about property in the U.S. that is a holdover from the days when one could move west and have land for the asking. Property values have shot up sky high in the suburbs so that large undeveloped tracts are only within the means of a developer who will get his money back—and then some—by subdividing. Thus it happens that in the postage stamp pattern of housing, landscape features—trees and the topography—are bulldozed into submission, and housing developments acquire an exposure of buildings and construction wounds on all sides—a sight that most people left the city to escape.

Nor are the money-lending institutions—notably banks and insurance companies—enthusiastic about backing unconventional land developments such as clustering, even though these approaches preserve the countryside. The element of risk in an imaginative plan induces most bankers to ride with the conventional scheme. For the most part, they won't support a developer who brings them a plan showing houses clustered to fit into natural terrain, allowing a small stream or a marsh to remain in place. They prefer to see a neat, geometric layout with lots and streets placed evenly, requiring the developer to flatten out the land and bury the flowing water in culverts.

Changing this trend will require a great deal more knowledge on the part of the American property owner. It is getting late to demand a role in developing the master plan for your community or changing a present one, but you should get involved. Remember that most people don't get involved, usually out of apathy. Your impact could be considerable.

⅗ If your town has zoning restrictions at all, do they fit in with a regional plan that has contingencies for each future plateau, as development pressures erode the local land base?

⅗ Does zoning encourage variation such as cluster housing? housing?

⅗ Are highways and streets taking only so much land as is minimally essential? As in the city, suburban streets take up an inordinate amount of room. The town of Radburn, New Jersey, nearly forty years ago coped with streets quite ingeniously by abandoning traditional zoning and creating thirty-five to fifty acre superblocks surrounded by main roads and entered by narrow dead-end lanes. Privacy was attained as houses could be built flush against these lanes with their courtyards behind. Children and pets were protected. Every house gained access to a central park through footpaths and bridges. And two great economies resulted: 25 percent less consumption of land for streets and a similar efficiency in laying utility lines.

"Stewardship"

The Open Space Institute in New York has produced some first rate documents (see Bibliography) that should convince the most conventional-minded developer or banker that imaginative use of open space not only provides more privacy but pays off in increased property values. The Institute, under the direction of ex-advertising man Charles E. Little, decided in 1964 to take an inventory of privately owned open space in the New York metropolitan region. Next, the landowners of large attractive lands were to be approached with the suggestion that they keep their green acres undeveloped in the public interest through various methods, ranging from making an outright gift to a conservation organization to giving the public a scenic easement. Land use regulations prevent a man from turning his property into a nuisance but they do not tell him to manage it in an enlightened way or prevent him from turning it into a housing development. "Therefore," Little recounts, "it seemed obvious that if open space was to be saved, some of it would have to be saved on a voluntary basis by the people who owned it. It was, and still is, clear that no government or combination of governments can ever raise taxes enough to cover the cost of purchasing even the minimal amount of public open space needed for recreation, amenity, and the maintenance of ecological systems in any growing metropolitan area."

The Institute published a manual for these private land-owners, entitled *Stewardship*. It is a masterpiece of persuasive logic. Not only did it explain the *values* of open space, but it presented a strong case to developers. The book contained examples showing the advantages of, for example, a cluster plan that sets aside 25 percent or more of the land for open space. One was done by the New Jersey firm of Halpern and Tuschak on a seventy-five-acre tract in Hillsborough, Somerset County, New Jersey. The developers convinced Somerset County planning authorities that a cluster plan would give the homeowner a more pleasant setting, would cost the county less in road, fire protection, and winter plowing services, and, most important, would allow connections to public water and sewerage instead of wells and septic tanks, which had been giving the county trouble. Standard zoning rules would have allowed seventy-two houses covering the entire parcel. But the developers instead built three clusters of twenty-four houses each, on half-acre lots.

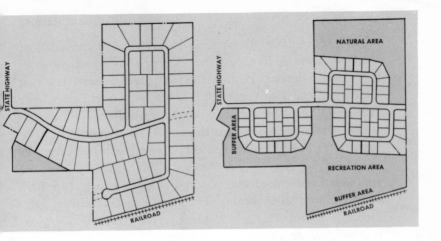

Conventional subdivision layout (above) contrasted to cluster plan actually used in Hillsborough, N.J., which preserved 40 acres of open space. (After a drawing in Stewardship, Open Spaces Action Institute.)

Thus forty acres were saved for a town park. Called the Village Green, the concept proved enormously popular.

Does your community have an open space or recreation plan?

Has land been designated for watershed protection and for wildlife habitats?

Zoning

Zoning *can* be valuable in protecting fragile environments. Many states have, for example, enacted zoning laws to protect coastal wetlands or floodplains. Both regions (see Chapter 8) are important natural breeding grounds, and are biologically most valuable to man. The Corps of Engineers, through regional offices, is most willing to donate to local governments information and maps on flood zones that should be kept free of development.

Some states have taken zoning decisions out of the hands of local governments to reduce the effect of commercial pressures and vested interests. Hawaii in 1961 passed a novel law that created a Land Use Commission. This body then divided the state into three classes of land—urban, agricultural, and conservation. Urban areas were for concentrations of people, agricultural for farm and garden production along with homes on lots of at least five acres, and conservation land was to be protected for natural resource or watershed purposes or managed for park recreation and scenic beauty. The Land Use Commission gathered its data for zoning by holding regular hearings in regions conveniently designated as conservation districts.

Whether Hawaii's or similar zoning plans succeed (and the Pacific Ocean state is now being strained considerably by development pressures) remains to be seen. Zoning reform, to be really effective, must be coupled with other changes, particularly in the tax structure.

One zoning proposal that holds promise is called *compensable* regulations. Although such a law has never been fully authorized, it has been much discussed and variations also have been suggested in Congress and at the state level. The concept was developed by University of Pennsylvania law professors Jan Z. Krasnowiecki and James C. N. Paul, and Mrs. Ann Louise Strong, of the University of Pennsylvania. As an urban consultant, Mrs. Strong was the author of one of the most useful and fundamental land use studies ever put together by the federal government (see Bibliography). Basically, the concept is based on state imposed zoning maps and regulations regarding the use of land. If a property owner chose to sell his land he would hold government guaranteed compensation for devaluation caused by the new regulations. In other words, he would be paid the difference between the property's value before it was zoned and the selling price, if it was less than the former. However, the guarantee on the property would be reduced after each transfer by the amount paid in compensation to the previous owner. Since the profit incentive would thus be taken out of land ownership, compensable regulations are a hard bite to swallow. But to some degree—perhaps in an even more absolute form—they are inevitable if open space is to be preserved in suburbia.

Taxes

Tax laws in this country have perhaps imposed the greatest restraint on enlightened land use, yet far too little has been done to reform the system. Since just about every local government finances its schools (and other services to a major extent) from property tax revenues, the incentive is to develop land, invite industries in, and put up new housing as fast as possible to fatten the tax base.

Only in recent years have some communities begun to realize the true results. There is a point at which land development becomes diseconomic, as I have already pointed out in discussing the growth crisis. In *Stewardship,* Charles Little noted that "As the metropolitan countryside fills up with houses, the people in them soon realize that the services they require cannot be borne by themselves alone. Often it takes a thousand dollars or more of locally raised tax money to educate a single child for a single year at a suburban school." Since each new home generally averages two schoolchildren, the costs of education force residents to attract industry to share the increased tax burden. And, Little wrote, "The cycle seems endless, because as factories and facilities reach into the suburbia, so new housing developments emerge as a necessary complement, restoring the tax imbalance that started the frenzy in the first place, and reducing the landscape to a shambles in the process."

In a recent work, *The Challenge of the Land* (see Bibliography), Little has refined his observations still further by presenting a solid case for towns to purchase open space to actually *save money in the long run* by avoiding the cyclical tax imbalance. He cites a 1965 controversy in Closter, New Jersey, to prove the point. Mayor James E. Carson wisely convinced townsfolk to purchase eighty acres of undeveloped land for limited recreation at a cost of around a half million dollars. If developed, the land would have been jammed tight with 160 houses producing 200 children (by one rule of thumb) that would cost $720 each or $144,000 in total to educate. Garbage services would cost $4,000 per year, police patrols, $6,000, fire hydrants, lighting, and other services, $2,000. These were costs that could not be made up without higher taxes throughout the community. As Little, William H. Whyte (see Bibliography), and others have pointed out, even when not counting the intangible aesthetic values of open fields and groves of trees, the preservation of *some* open space (neither of these men advocates exclusivity at the expense of the congested cities) is economically sound in case after case.

The City of Mercer Island, a rapidly developed enclave on Lake Washington, only ten minutes or so from downtown Seattle, has been pondering an unusual move to save open space that had not been resolved as this was written. In three decades, after it was joined to Seattle by a floating bridge, Mercer Island grew like Topsy. Its verdant forests, lush ravines along the lake, and ridge-line meadows were developed in a most haphazard and disorderly manner. Now the town is considering "A Proposal for Planned Saturation of Mercer Island," that would create a park out of 670 acres of the remaining 1,000 acres of undeveloped land, "and by thus limiting future home building, reduce anticipated capital outlays and operating costs for schools and city." The proposal has been carefully analyzed with respect to zoning and tax side-effects. The city has taken advantage of a federal "701" planning grant to perform its cost-benefit study. While it was estimated that the 670 acres would cost up to $15 million, compared to the city's bonding capacity of about $2 million, it was recommended by city

manager Donald Hitchman that various known methods of
open space acquisition as well as innovative approaches be
thoroughly explored. One new thought was the concept of
"trading dollars," whereby a community anticipates the costs
of capital improvements and services for new development and
then sets aside this money for open space (*i.e., to prevent a
development imbalance*).

There are a number of proposals for tax reform. One good
one is the suggestion that the federal income tax be amended to
allow a portion of capital gains taxes on land sales to go into
federal acquisition of open space or be paid back to the towns
for the same purpose. Either way the public would gain protection and additional leverage in planning land use. Howard
County, Maryland, has created an open space fund by putting
a portion of the real estate transfer tax into acquiring undeveloped land for public use. The Federal Housing Administration
has estimated that on a nationwide basis this measure could
provide a million acres of new parks in the next thirty-five years.

There is federal enabling legislation, and the states too have
laws, granting property owners tax advantages for providing
"scenic easements" or public access to land that will be kept in
its natural state. Taxation is also relaxed to encourage the preservation of historic sites and buildings. In fact, the Department
of Housing and Urban Development provides 50 percent
matching grants to save historic properties.

Land Banks

Looking at present trends, it does not appear that even a
revolution in zoning and taxation will give communities sufficient leverage to plan ahead for their land. The most effective
potential tool to round out their planning arsenal would be
legislation to enable large-scale public acquisition of land. To
be sure, local bond issues have for years been one way of raising

money to buy park land. But seldom do these moves take place in time to get a bargain on the deal. Towns in the European countries and Great Britain have for years, however, been buying land beyond the fringes of spreading urbanization well before growth pressures have forced prices up beyond public reach. When a city does this it gains the power to regulate its development in an orderly and efficient manner. When it decides to locate schools, transit lines, and parks, it can use its own land and thus prevent wild speculation or indiscriminate development. The result, depending on how much land is publicly owned, can be balanced growth with a stimulating mix of features—schools, shopping centers, parks, housing areas, and so forth. The city can sell land to private developers with stipulations as to its use and at a reasonable—not inflated—price that will tend to hold other land values in check. Presently there is legislation pending in Congress that would let the federal government assist the establishment of land banks throughout the country to acquire public land for long-term development. Later in this chapter new towns will be discussed in relation to this concept. It suffices here to note that an obstacle to beginning a new town has been high initial land costs. Buying or leasing land from land banks for new communities would help.

Already, some cities and states have moved toward innovative financing for public land acquisition. Baltimore, for example, has a revolving trust fund for buying land ahead of need, and Pittsburgh has set up a land reserve fund to minimize the widespread effects of land inflation.

New Towns

Indeed, the most promising prospect for bringing order *out* of the suburbs if not into them is the so-called "new towns" concept. There are different versions and even examples of what a new town should be. Some planners see it as a satellite suburb joined to an urban hub by a transit line. Others conceive of it as a completely self-contained city outside of the city, able to hold its population by providing industry, corporate branches, shopping plazas, theaters, and other services. Each will qualify, provided it is planned from the very first building block as an integrated unit. But the ideal new town, in the U.S. at least, would appear to be a compromise, a well-ordered new community that is a bridge between urban and rural experience, providing the most attractive elements of each world. The new town's compact villages, apartment and service complexes, parks and plazas, can generate the excitement of the city—a concentration of different types of people all "doing their thing," hustling from place to place, mingling and talking with one another, walking dogs and children. But the new town's interweaving of open spaces, its provisions for some single family housing in carefully landscaped terrain, its manmade lakes and groves of trees offer the stimulation of nature and the serenity of the country.

The new town idea is really rooted in a life style that western man has aspired to, if not attained, since his very beginnings.

The Greeks can be credited with establishing the notion that the environmental amenities of cultural and athletic facilities coupled with orderly city planning would provide an atmosphere in which the talents of man would flourish in splendid harmony. In our time, the Englishman Ebenezer Howard laid down the basic premises for new towns in his writings about "garden cities" at the turn of the century. He proposed a town of thirty thousand people in which the vibrancy of the city would be enhanced by the natural beauty of the countryside. There would be industry, commerce, and culture thriving in the best of settings, laced with parks and greenery.

Countries in Europe and Scandinavia and Great Britain herself have gained a long headstart on the U.S. In each country, new towns have been built differently. For example, Sweden was a rural nation until after World War II. But since 1900 the government had been acquiring land to anticipate urbanization. Thus residential satellites with the character and orderliness of a model community have been built around the city (i.e., Stockholm), connected by transit and well-provided with public services. England, though, developed new towns more on Howard's precept since the country was overwhelmingly industrial and urbanized over a century ago. Her new towns are more self-contained, not as purely residential as Sweden's. By consensus, scholars find Tapiola, Finland, the most exciting new town mainly because of its healthy social character—a good mix of all kinds of people affording housing at a wide range of costs. It is spectacularly landscaped throughout its 670 acres, but holds comfortably a population of seventeen thousand (twenty-six persons per acre) and provides jobs for six thousand.

In this country, Radburn, New Jersey, was the first community hailed as a new town. As noted previously, this is where the automobile was put in its place so that residents could have more open space and privacy. But Radburn, for all its planning brilliance, is still mainly a pleasant suburb, not a new town by Howard's or our guidelines today.

Reston and Columbia, both outside of Washington, D.C., come much closer to the ideal. While upwards of sixty new towns are now being planned or built in the U.S., Reston and Columbia are the farthest along. The former, eighteen miles out of the District of Columbia in rolling wooded Virginia countryside, plans to have seventy-five thousand people in four or more villages on the 6,800 acre site by 1980. The villages and a main center will take up two hundred acres, industry and government agencies will use nine hundred acres, and planned for public use are over three thousand acres or 42 percent of the land. Reston's present or planned features include man-made lakes, walking, riding, and bicycling trails, and common gardens. The present population is 8,500 and the industrial park employs 2,100.

Columbia, in Maryland, is not so far along. Planned for 125,000 residents, it has provided more low cost housing than Reston, seems less precious architecturally and less of a white, upper middle-class ghetto, as Reston is sometimes called.

The first federally subsidized new town is Jonathan, Minnesota, a five thousand acre site in hill country twenty miles southwest of Minneapolis. It has been launched with a HUD guarantee of up to $21 million to pay off debts incurred in the initial acquiring and developing of land. Government backing in the early going is critical, for so far lack of such support has hindered new towns and forced some, like Reston, to make compromises. Jonathan plans five villages, each holding up to seven thousand people, and a high-density center of ten thousand. Over twelve hundred acres are reserved for industry, nine hundred for recreation and open space.

While new towns are the wave of the future, strong government backing has not yet developed. There are new town proposals pending. Basically they provide subsidies to cover developers' initial, pre-sale-period borrowing costs, enabling legislation to acquire land, and assistance for infrastructure construction (i.e., public facilities). A vigorous new town program is long overdue. Aside from providing communities with a good deal of natural land as open space (on the average 25 percent, more than twice that of cities), the new towns are opportunities in land-ethic planning. Because they are planned as a whole, their environmental assets can be both preserved and effectively utilized. They are the ideal testing grounds for new transportation systems, new housing materials, and new construction methods. They will not replace cities but they will give the cities a new chance to regenerate their air and water and recover open space. The new towns could halt the aimless spread of the city outward and absorb much of the rural migration inward. In addition to all of this, the new towns are examples that can be followed in all suburban developments— even though new town balance cannot be achieved. Since government support will be based on high planning standards, the new towns ought to be showcases of land management. As a citizen, you are not in a position alone to swing a new town into place, but you are able to observe the development guidelines apparent in the new towns. For example, in a new town, you should not see houses put down on steep slopes so that massive erosion can result. You should not see more concrete and asphalt than grass and wooded areas. You should see people walking everywhere—a rare sight in most conventional suburban enclaves. There is plenty of room for micro-new towns throughout the U.S. and land for more new towns too.

RURAL AMERICA: PROBLEMS

The familiar comment of Thomas Jefferson that "Those who labor in the earth are the chosen people of God. . . ." has no credence in America three hundred years later. We cleared the forest for the farm and now we subdivide the farm to make way for more homes pushing out of the city while the farmer migrates back to the city looking for whatever work he can find.

Urban land must be used more efficiently. Suburbia is the proving ground of new concepts in land use. Out in rural America lies a different but equally grand opportunity. Historic and generational ties are less disturbed. Towns that to us today seem as logically planned as they are picturesque have been less uprooted than metropolitan villages, although the country landscape is fast being made ugly. In short, it is in the country where

we have a chance to make "new" towns by revitalizing and
preserving the best of the old. In the city and in suburbia,
simply restoring the old structures won't meet the high-powered,
streamlined life style and movement of these regions. Not so in
the country. It is where a person can—and should be able to—
feel the force of the land and the climate. Neglected towns lie
waiting to be uplifted from the straits of bowing down to tourist
travelers or presiding over the slow death of the small farmer.
The Department of Agriculture administers upwards of five
billion dollars in subsidies to pay farmers not to farm or to
buoy them up artificially. This money could far more effectively
be spent in programs to rejuvenate rural America, keeping the
farmer down at the farm doing other things besides farming, in
short, participating in a rural renaissance.

Out there in rural America lie the scattered remnants of
tradition, native craftsmanship, and reverence for the moods of
the land. If these could be sustained, the nation would benefit
enormously. But rural America is in trouble. Industries—
notably the mines—have plundered the face of it, leaving
mountains decapitated, hillsides scarred, and eroding gullies
down which sulfuric wastes still run. Dying agricultural practices and the consolidation of small farms have driven rural
inhabitants to the city and, finally, caught in the path of urban-
suburban expansion, farmers can no longer pay the property
taxes on lands now assessed at "fair market value" for residen-
tial-commercial subdivision.

Typical is the scene in Suffolk County, Long Island, still New
York's largest agricultural district and where seventy-one
thousand acres are devoted to potato, vegetable, duck, and dairy
farms. Yet in two decades the number of farms has dropped
dramatically from 2,187 to 1,020 and at the rate the county is
expected to grow—from 1.2 million to 2 million people—by
1985, barely six thousand acres will be under the plow. There
have been some attempts to save the farmland by zoning it for

open space, but land speculation has made it a losing battle. In 1960 an acre might sell for $2,000, high enough then to undermine the land's worth in agriculture, but now prices fly upwards of $5,000 an acre. It pays to sell it off and invest the capital, especially since taxes have doubled per acre up to $40 or more. (In Delaware the taxes are still as low as $2.50 on farmland.) Market fluctuations have made potato farming marginal. All in all, there is little reason to keep the land from development. A $25 million bond has been proposed to acquire open space for the future, but nobody seems to care much about it and such a measure may never be on a ballot.

The same picture can be painted all over rural America. Where she has been raped—as in Appalachia—by strip mining or over farming, there is such desolation as to make restoration of environmental quality a dim hope. In Pennsylvania and West Virginia, for example, not only does acid mine drainage continue to contaminate thousands of miles of streams (see Chapter 1) but the land is dangerously undermined by old diggings. It is so honeycombed that the ground slumps continually in cities and suburbs. Poisonous fumes emanate from below, and mountains of coal tailings—called bulm heaps—are smouldering from spontaneous combustion, polluting the air. States (e.g., Kentucky) are only beginning to act to reclaim past degradations and to insist that new operations protect the environment as they go. Only Pennsylvania makes a mine liable for surface damage caused by underground collapse. And more states should require natural resource extractors to post performance bonds before mining. Alaska passed a good law under Walter Hickel, then Governor. Natural resource companies could not take minerals from the ground unless they established processing plants in the state. This would benefit regional economies but, more important, force the industries to hold a stake in local environmental protection while giving the state and local governments some leverage over mining operations.

It was no longer a case of "cut out and get out," to use a phrase applied to early lumbering.

Agricultural policies (see Chapter 9) have encouraged rural land practices that have abused the soil and tax policies have forced the farmer to yield to the developer. Now around the cities, up to three hundred miles distant, rural land is menaced by the leisure-time migration—the search of increasingly af- fluent Americans for a "second home." Dennis Durden, geog- rapher and planner, calls this invasion the "new colonialism."

Nearly every state feels the effects of the leisure settlement wave. In our proposals we'll suggest how the movement could actually enhance worn out or desolate environments. But more typical is its threat to prime wilderness. In Virginia there is a rush for the hills and valleys of the Shenandoah—an easy day trip from Washington, Richmond, or Charlottesville. The weekend papers are full of advertisements for vacation develop- ments in the land that Lee's generals used to roam.

Of course rural land is sacrificed and, while people will be happy at first in these A-frame communities, one wonders what the long-term effect will be. "Back to Nature is New Trend" ran the headline of one ad for a place dubbed "Indian Acres" south of Fredericksburg, Virginia. Quite possibly the developer's plans to divide the tract into "funstead" sites for "camper, rec. vehicle, or tent" can be accommodated without ruining the place, but it would appear unlikely that this countryside will continue thus to provide unspoiled "the oldest pleasures nature has to offer."

Many areas attractive for the new colonists have been blighted by the failure of an agricultural livelihood and the "out- migration" of youths coupled with "under-employment" (avail- able jobs of low wages). So the recreation industry offers great promise and is impossible to resist in these rural outposts. But more important in the long run are rural environments' potential as communities restored to all-season health. If that were the case they would be even more attractive as resort colonies because they would give off an indigenous flavor—the glow and pride associated with a good and supportable way of life. And these communities would not have to stoop or compromise one bit to please and provide for the colonists.

So rural America is barely hanging on, buffeted by migrations of exploiters, the exploited and vacationers going in opposite directions—and none of them are involved in the life and future of the citizenry that have lived in these regions for decades and taken strength from rural traditions.

RURAL AMERICA: PROPOSALS

The Farmers Home Administration provides credit to communities of under 5,500 people for recreation facilities and waste disposal, and other programs assist towns in landscaping, soil conservation, and planning. However, while the federal government exerts an overpowering influence on rural life through farm subsidies, parks, flood control, hydroelectric projects, and other general activities, it seems to care little about the consequences and does little or nothing to help small towns pick themselves up. In short, present federal interest in rural America —and for that matter, many state programs are just as guilty—does not value the dignity and self-respect of the rural inhabitant.

Rural communities can get started toward what could be called the rural renaissance by examining the qualities that make them so attractive to themselves and outsiders. If they keep asking themselves what is important about their environment, the chances are that they will feel very strongly about protecting open space and retaining simple architectural charms. They will likely conclude that long-range planning is an imperative to avoid ugly, commercial intrusions. The evidence indicates quite plainly that "sleepy" rural villages prosper as communities by retaining their tranquil, rustic airs. While individuals may get rich selling out to resort or commercial developers, there is too often nothing but collective misery because the community's beautiful face is fast scarred, and revenues from temporary or seasonal business are of questionable benefit.

It is proposed here not only that rural revitalization become the major, consuming goal of the Department of Agriculture but that rural communities take part in regional commissions or planning bodies (as has been done in some areas) to draft guidelines for land use.

One model for such regional planning is the Ten-County Union in south central Iowa that was formed to pool resources, to impose order on growth, to anticipate local economic shifts, to attract business that would be compatible with the environment, and to conserve open space. "Tenco" is reported to have achieved success in all these objectives.

Another good approach is Wisconsin's outdoor recreation plan instigated by Senator Gaylord Nelson, as Governor, and based on an extensive land inventory that took into account natural resources, watershed, and conservation values. Where these overlapped, the state mapped out parks and wildlife refuges to be set aside somehow, or acquired outright through bond issues.

A strong regional organization could have prevented chaos and speculation in Weston, Illinois, where the Atomic Energy Commission is building a $300 million proton accelerator. Here was a prime chance for a community to gain control of land and build a new town. The University of Illinois School of Architecture warned that failure to adopt new zoning and land use controls, as well as a master plan, would be disastrous. But Weston just couldn't get together to hold off uncontrolled development.

Every regional planning body should follow the latest example of the Tennessee Valley Authority (which isn't always exemplary) in hiring an ecologist to oversee one of its many watershed projects. Ecologists can tell planning councils how their best notions may have destructive side-effects. An ecologist can pull together a spectrum of environmental data—biological, geological, hydrological, and climatic—and can forewarn of poor land development practices of the sort that produce annual tragedies in the Los Angeles suburbs. In January, 1969, torrential rains loosened the surfaces of these hills so that they slid down their slippery linings of clay strata, carrying fifty thousand homes, many to total destruction. Ecologists can tell planners how to weigh the effects of air currents, water drainage, flood control, erosion, plant and animal interactions, and open space —in sum, an environment's biotic criteria. This sort of planning is vital in rural America, where there is so little protection against ecological abuse. How do you obtain such advice? Inquire of local conservation organizations and write national groups such as Open Space Institute to find out what has been done in your area in the way of natural resource planning. Of course college architectural and planning departments should also help.

Rural towns should also look into the potential of reviving native arts and crafts—from furniture making to quiltwork. Such products are coveted these days, as they have become a memory of the past. Small villages throughout the nation have been quite successful in a handicraft revival. In New England, traditional wares are sold through country stores—usually by mail order. In Monterey, Virginia, a pretty hill town in sheep and cattle country, where trout and maple sugar run too, citizens set up a crafts association to encourage the "cottage industries." It held classes in woodworking, weaving, painting, and pottery, rented a workshop and selling space, held a festival, and now seems to be prospering. In this depopulating county where farming has waned, the jobless and the older people were rallied and, it is not trite to add, the town could take pride that its tourist products were native and not made in Japan. Expanding on this theme, in a civilization that has become overly homogenous (from television programming to fashion), it would certainly pay to offer incentives for local industries to provide more basic goods and services. While local manufacturing of automobiles obviously would be uneconomical, the production of food and clothing would profit from indigenous ingredients, materials and skills, and regional pride would be enhanced greatly. Instead of sending all their best beef cattle to mid-west packers, southwest states would have local meat industries. The fish and lobsters in Maine would benefit that economically depressed state a great deal more if processed in Maine. This approach, however, will need incentives and perhaps subsidies to become economically feasible.

There is not enough that can be written about the benefits from the sort of community planning that sweeps up a whole town. To be sure, such work can be dull. It means attending town and county meetings, preparing for hearings on complex confrontations. It is time-consuming, often entailing door-to-door rounds of the neighborhood and subjecting yourself to ridicule or malicious behind-the-back charges. But it always pays off. A fine example of such civic-mindedness has been going on in Bayfield, Wisconsin, a picturesque town on the shores of Lake Superior near the glorious Apostle Islands. Lumber, farming, and fishing had given way to tourism, but the surge of visitors threatened to destroy Bayfield's character, clutter its waterfront with marina activity, and mar its narrow streets and approaches with food and souvenir stands, a jumble of signs and flashing neon lights. Under a federal education grant, administered by the University of Wisconsin Department of Landscape Architecture, Bayfield's citizens joined in an all out spruce-up coupled with the setting of design and land use controls that were presented in a first-rate eighty page "Blueprint for Bayfield." That masterplan looks well into the future, but it is also a comprehensive analysis of Bayfield's present vital assets. If every rural town is unable to benefit from such a study, at least school students could make a project out of inventorying the important features of their town environment.

Open space zoning is a helpful solution to protect rural land but, as previously pointed out, it should not be relied on as a permanent defense. It is strengthened considerably if it is encouraged by local property tax incentives, i.e., taxed as agricultural land not on the basis of its fair market value. Another approach is the scenic easement, already mentioned as applicable in suburbia. It is an excellent solution in rural America where the main consideration is keeping forest and farmland preserved to meet future demands. For years, state game departments have purchased hunting and fishing rights on private lands or streams, but such easements do not restrict an owner's use of his property. A scenic easement does not necessarily grant public access but it does regulate the way land is used. It may be purchased to preserve farmland as a buffer strip to a national park or wilderness area. The easement may flank a highway and thus be useful to prevent honky-tonk strip development. It may be privately owned shoreline valuable just as a wildlife habitat or a scenic stretch in a region of overall beauty. Except in congested areas where development pressures are terrific, scenic easements are generally inexpensive, often around $20 an acre. In rural America they would serve now as a means of controlling future land use.

If towns keep close track of what various government agencies are up to, they can also occasionally latch on to an open space grant for what is known as multipurpose land acquisition. This in short is the purchase of land adjoining a public project —a highway, school, or park, for example—that will cost far less if it is included in the original transaction. But if not caught in time for that, such land is often priced out of reach because of its proximity to the new development.

CITIZEN ACTION

A chapter on land use in the United States cannot conclude without at least a summary reference to citizen-action and conservation organizations that for many years have been setting enlightened precedents in land acquisition and management. In addition, nonprofit educational operations such as the Washington-based Conservation Foundation have provided research and models that are most useful to practical community planning. The latter's study project on Rookery Bay, Florida, produced a plan that, if followed, will accommodate both urban development and conservation in the growing estuarine community of Naples.

In the past decade, Nature Conservancy has raised enough money, often hitched to matching grants, to acquire nearly 190,000 acres in forty-one states, turning this land over to parks, refuges, or wilderness areas. The Conservancy has been dramatically successful in saving habitats of endangered wildlife, islands (e.g. off Maine and Georgia), and coastal wetlands. The organization's performance has won it a six million dollar line of credit from the Ford Foundation.

In the metropolitan area, Philadelphia Conservationist, Inc., has, since 1954, purchased nearly ten thousand acres of conservation land, beginning with the Tinicum Waterfowl Marsh, a 245-acre station for migratory waterfowl and shorebirds. The nonprofit group promotes conservation throughout the Delaware Valley, has set up a watershed management association among other projects, and has protected land mostly within one hundred miles of the city.

In Connecticut, the "land trust" concept has taken root. Now there are more than thirty-five of these private, nonprofit corporations set up to take money or land which they in turn put into conservation parcels. These are usually turned over to an appropriate government agency with stipulations as to how they are managed. These small private groups have been able to move quickly when a piece of wild land was threatened by a development and the trusts appear to have gained landowners' confidence as no public agency can.

Most of the New England states for some time have had enabling legislation allowing towns to set up conservation commissions. The first state in the act, and so far the most successful, was Massachusetts. It now has nearly three hundred commissions, which are boards of citizens who protect natural resources, come up with land use suggestions such as wetland or floodplain zoning, and generally try to keep both residential and industrial developments within strict environmental limits. In Connecticut, the commissions initiated state legislation giving tax advantages for preserving farmland or open space. In New Jersey the commissions are represented on town planning boards and, in this the nation's most heavily industrialized state, the movement has been an amazing success.

A fat directory is needed to catalogue the conservation organizations. In an appendix, you will find some of the main national groups listed, along with a description of their activities.

Over the past two years at least, surveys have revealed that an increasing, and already large number of Americans yearn for the experience of nature, countryside, the sound of rushing water, rustling trees, and birds singing. Those who have traveled around Europe, where the line between parks and cities is not so sharply drawn as in the U.S., are embarrassed when it is their turn to steer a foreign visitor into this nation's beauty spots. Indeed, spectacular wonders abound, but the trail to them is sad and littered. One is tempted to justify the U.S. landscape by thinking back on history. We are a melting pot and there are no common roots or traditions. We have no compelling reverence for old forms or styles. And during our development there was the constant excitement of the industrial-technological revolution in the east, and the discovery of the frontier to the west. When we finally realized what a mess we had made, there was no open space left to speak of in the cities and the suburbs were spreading out of control. We created great preserves and called them parks. But they were so far out of our experience, cars were necessary to reach them, and so cars clog up the parks.

It will never do. We may as well face up to the fact. The oldest large democracy has been unable to get a grip on its expansion and development. It has little to do, either, with historical trends. It does have to do with our Constitution and tax laws, which from the very beginning have encouraged development of the land.

Congressman Paul McCloskey, who wrote the preface to this book, suggests that only by developing a national land use policy with the power to implement zoning controls can we reverse the trends. Certainly there are precedents. In 1967, Great Britain set up a land commission to purchase land for future growth and reduce the incentives to land speculation.

Those, including McCloskey, who have discussed a land use commission in the U.S., envision a similar national zoning plan designating the best uses of the land. If this plan devalued a piece of property, its owner would be reimbursed from a trust fund. If his land went up in value, the increase would be imposed against the property as an assessment lien and paid to the land commission trust the moment it was sold. This scheme is very similar to the compensable zoning discussed earlier. To most Americans, such thinking may appear very radical. Certainly new measures to cut down land speculation and reduce the pressures to develop land, as well as ways of returning land profits to communities, are inevitable sooner or later.

8 ✤ THE SHORELINE: MARSHES, WETLANDS, AND ESTUARIES

ONE THIN STRIP of our environment is of the greatest concern to ecologists. This crisis corridor is the shoreline that borders the nation along the sea and through the Great Lakes. It is the fragile zone where land meets water, where rivers flow through estuaries into the ocean, where marshes, delta islands, and tidal flats teem with wildlife. The most productive part of the environment biologically, and perhaps the most complex ecologically, the shoreline is where all of the pressures on the environment have concentrated most forcefully, and where they constitute the greatest threat. Thus the shoreline is the only ecological zone which will be discussed separately in this book.

One third of the U.S. population (and nearly half of the urban population) lives in counties along the coast. Forty percent of the nation's manufacturing plants are found in these counties. From 1930 to 1960, the population of estuarine areas grew 78 percent, compared to a national growth of 46 percent. At the present rate, by the end of the century there will be nearly two people per foot of waterfront.

Wherever the shore is penetrated by natural harbors or deep coves, you will find urban and industrial clusters or plans for additional development and even new settlements—for marinas, vacation housing, oil refineries, petrochemical plants, steel and aluminum mills, and electric power installations. Cities such

as Boston, New York, Chicago, Los Angeles, San Francisco, and
Portland, Oregon, already have built or considered airport
facilities that consume valuable waterfrontage and marshland.
Everywhere the shoreline is threatened with massive encroach-
ment from the land, yet no less dire is the threat from seaward.
Vessel discharges and spills, pollutants formed in offshore
sludge and waste dumps, and noxious chemicals that have run
off fields and dumps all swirl back into the coastal or lakeshore
waters with disastrous effect.

WHY THE SHORELINES
ARE SO VALUABLE

As a piece of real estate confined to a band 88,633 miles long
—99,613 counting the Great Lakes frontage—the shoreline
obviously commands a premium. Industry needs water access for
shipping and water for cooling, and everyone likes water if for
no more than the boundless view it affords. What may not seem
so obvious, except to those who depend on it, is the value of
coastal shallows to the nation's—and the world's—fisheries.
Seven out of ten fishes in the U.S. commercial catch need these
waters at some stage in life. The shellfish industry is entirely
sustained by the shoreline environment, and oyster and clam beds
and areas rich in crabs are being closed down or wiped out
by pollution and habitat changes at an alarming rate. Why is the
shallow water near land so different?

Nurseries and Feeding Areas

Estuaries are defined as waters where fresh streams and rivers
meet and mix with salty seas. Wetlands are the shallow shelves
extending beyond the shore into both fresh and salt water.

Marshes are the shallowest wetlands—inland and along the shore—although they include small islands and patches of vegetation on ground above water. According to Dr. Eugene Odum, University of Georgia ecologist, a marsh can produce twenty times as much food or organic material as the deep sea, seven times as much as an alfalfa field, and twice as much as a cornfield. The reason for this is that mineral-rich seawater, carrying organic substances that have decayed underwater, mixes with the topsoil runoff and countless other land-originated nutrients to produce a mighty harvest. The key to this production is the solar energy that penetrates the shallow depths and unlocks the process of photosynthetic conversion, hastening biological decomposition and the growth of new matter and food. Hence, the productivity of the oceans and lakes is sharply curtailed somewhere between one hundred and two hundred feet, the range of sunlight depending on the turbidity of the water, and at depths greater than four hundred feet there is virtually no plant growth. Dr. Stanley Cain, a distinguished ecologist, summed all this up quite beautifully when as Assistant Secretary of Interior for Fish, Wildlife and Parks he testified before Congressman John Dingell's Fisheries and Wildlife Conservation subcommittee in 1967. The tidal cycles blend the most fertile elements of the land and sea, he said, and, "Thus a sort of constantly stirred rich broth is provided in a sheltered environment for small and microscopic plant and animal plankton to form the abundant food for successively higher links in the food chains that make up a web of life. The result is phenomenal," he continued. "Some single estuaries are the spawning grounds, nurseries, or growing up places for two dozen or more species of commercially important shellfish, crustacea, and finfishes."

Anadromous (migratory) fish are sustained by coastal waters as they pass through to spawn in freshwater streams and rivers. Their young—salmon, some trout, shad, striped bass, alewives, and other species—come back and grow up in estuaries where they find the food they must have as well as security from offshore predators. The winter flounder spawns in the estuaries. Bluefish larvae hatch at sea and then swim to the nursery, and eels spawned in the ocean return to live there. The coastal shallows are crucial for crab, oysters, and shrimp. Thus preservation of coastal habitats is a biological necessity.

Minerals, Wildlife, and Recreation

Oil, natural gas, sulfur, and construction material are sought along the offshore shelf, often in proximity to estuaries and wetlands. Huge deposits of phosphate rock have been found in Tampa Bay and Pamlico Sound and even the nation's only subtropical park, the Everglades, is threatened. Mobil Oil Company has just leased over 100,000 acres of land where some oil was discovered on Indian reservations just north of the park. This area is critical as a watershed for the Everglades estuary. Already oil rigs have sprouted in the Gulf of Mexico, off Santa Barbara and in Cook Inlet, where spillage poses a hazard to coastal fisheries and wildlife. Just who gets first priority in tapping the resources of the Continental Shelf— mineral extractors or fishermen—is a question being debated in the highest government councils, since too frequently these interests are in conflict.

Valuable fur-bearing animals, such as seal, mink, muskrat, otters, and beaver, live in marshes, wetlands, and estuaries. In fact, nearly two thirds of the nation's fur sales alone come from animals inhabiting Louisiana's vast delta network. But the coastal region is far more notable for its importance to migratory waterfowl and shorebirds. Geese, rails, snipe, and duck spend the winter in the Atlantic, Gulf, and Pacific estuaries. Year-around inhabitants include waders such as the heron, crane, egret, and ibis; birds of prey such as the eagle, hawk, and osprey, and waterfowl such as the gull, tern, cormorant, loon, grebe, and pelican.

In 1962, the outdoor Recreation Resources Review Commission, a presidential task force, concluded that for public enjoyment alone there was an urgent need to acquire and protect far more of the U.S. shoreline. Excluding Alaska, the commission found that 21,724 miles of waterfront in twenty-eight states (including those on the Great Lakes) were suited to recreational activities. At that time, however, only 1,209 miles were owned by or open to the public. Since then, through the Wildlife Restoration Act and the Land and Water Conservation Fund, federal agencies and states have cooperated in acquiring more public shoreline. Eight national seashores have been set aside and others proposed. Some states have acted directly to conserve or buy coast land and their programs will be described later in this chapter. Scant progress has been made, however, toward what will be needed for public enjoyment even a few decades from now. It is unlikely that the waterfront will have any recreation value if it is polluted or jammed tightly with developments. This is happening at a dizzying pace. In an exchange with Harvard students at an oceanography conference in 1969, Captain Jacques-Yves Cousteau, author and director of the Musée Oceanographique of Monaco, lamented that "There is no place where pollution is not acute but, of course, in coastal areas it is more so. Just off some very popular jet-set types of beaches, the pollution is so bad that, if a swimmer drinks about half a glass of seawater, he has a chance out of five to get hepatitis."

One can compute the value of the shoreline environment in terms of fisheries, mineral wealth, wildlife, and recreation through endless illustrations, but it is difficult to pin down figures. How do you assess the view that attracts so many people to work in San Francisco? What is it worth to know you can drive to a stretch of unspoiled beach and walk alone along the sea? A 1969 Report, "The National Estuarine Pollution Study," by the FWQA, noted that sport fishermen spend up to one billion dollars a year and that because of income generated through fisheries and recreation and port activities the average personal income in coastal counties was $500 greater than else-where in the U.S. With some additional extrapolating, the study figured that the total direct economic benefit of the estua-rine zone (in a healthy condition) was about $60 billion. Just as effective a summary of the value of marshes, wetlands, and estuaries were these words in the FWQA report:

The great unique use of the estuarine zone, which makes it of primary importance to man and his civilization, is its place in the life cycle of many animals which aid in converting solar energy into more usuable forms. While no life form can be singled out as irreplaceable, the kinds of life which need the estuarine zone to survive represent essential links in the energy conversion chain upon which man depends for survival. Many of the human uses of the estuarine zone depend directly or indirectly on the existence of the estuarine zone as a healthy habitat.

WHAT ARE THE THREATS?

Figures vary, but it is generally estimated that there are about 26.6 million acres of estuaries in the U.S., of which about eight million acres are considered to be important fish and wildlife habitats. Nearly 600,000 acres of the most valuable re-gions have been destroyed by dredging and filling, let alone degradation by pollution. Here is a table showing the status of the largest estuarine states in 1967:

ACREAGE

STATE	TOTAL	HABITATS	LOST	PERCENT AGE LOST
Alaska	11,022,800	573,800	1,100	.2
Louisiana	3,545,100	2,076,900	65,400	3.1
North Carolina	2,206,600	793,700	8,000	1.0
Virginia	1,670,000	428,100	2,400	.6
Maryland	1,406,100	376,300	1,000	.3
Texas	1,344,000	828,100	63,100	8.2
Florida	1,051,200	796,200	59,700	7.5
New Jersey	778,400	411,300	53,900	13.1
California	552,100	381,900	255,800	67.0
Alabama	530,000	132,800	2,000	1.5

This table gives you those states with more than a half million acres of estuarine land. But states like New York, New Hampshire, and Connecticut have suffered high percentages of damage to their precious marshes. Since this table was presented at a Congressional hearing, state action has varied. Maryland and Virginia have braced themselves in vain against a terrific onslaught of development in the Chesapeake, whereas California has gained a reprieve against landfilling in San Francisco Bay that accounted for 192,000 acres lost.

Landfill and dredging projects do indeed appear to constitute the biggest encroachments on shoreline. Invariably, these activities not only occupy estuarine acreage but they produce a heavy volume of wastes and other pollutants. Landfill requires dikes and bulkheads to provide stability and prevent erosion. Thus the flushing action of the tides, the blending of freshwater runoff with saltwater elements, is prevented. Dredging disrupts bottom food chains or suffocates animal and plant organisms in silt which is churned up by the movement of dredge scoops and then is dumped back in the water to one side of the deepened channel or cut. A research paper by the League of Women Voters (see Bibliography) notes that "Because the estuarine habitat is complex, tampering with it leads to unanticipated side-effects. *No estuarine area, once destroyed, has been successfully restored."* (Emphasis added.)

Dredging may well destroy the oyster industry in Galveston Bay, Texas, if pollution from industries along the Houston Ship Channel doesn't contaminate or kill all life in the estuary first. The oyster shells that have accumulated for ages are ideal material for concrete aggregate and road surfaces, and are used in making lime, various chemicals, and chicken feed. In a state where political back-scratching at public expense is the rule rather than exception, the Texas Parks and Wildlife Department, a resource protection agency, was unable to resist making money (thirteen to fifteen cents a cubic yard) from shells dredged from the bay bottom by private contractors. Investigating this activity for an article in the *National Audubon Society Magazine* (Nov/Dec, 1968), naturalist and author George Laycock discovered that dredgers were creating havoc. If their destructive machines did not tear into live oyster reefs they had the same effect by smothering the beds in silt and debris.

Along every coast, dredges are at work, scooping out sand and gravel for construction, making navigational channels, or lifting the bottom for material to fill marshes and wetlands. Much of this activity is conducted under the auspices of the U.S. Army Corps of Engineers, which issues permits for both dredge and landfill operations. The National Sand and Gravel Association testified openly at a Congressional hearing that much of the forty-two billion tons of construction material estimated to be needed over the next thirty years would come from estuaries near large cities (i.e., the most valuable coastal shallows in Long Island Sound, the Delaware River, the Chesapeake Bay, Mobile Bay, and other inlets along the Gulf, to name just some).

Other kinds of public works projects are just as destructive. Some communities put dikes around marshes or cut canals through them to prevent mosquitoes from breeding. These areas are either dried up or turned into freshwater lagoons, their natural value altered or irreplaceably destroyed. All too often, if a road has to be built through a marsh, it is deemed too expensive to construct a bridge on piers that would allow water currents to move unobstructed through the wetland. Instead, the engineers interdict the essential waterflow by building causeways, which turn the marsh into a stagnant pond collecting all kinds of pollutants. As already noted in Chapter 5, cities have used their estuaries for dumps. Dams and canals serving flood control projects, or water storage, alter estuarine ecology disastrously. Not only has the nation's only subtropical estuary, the Everglades, been slowly starved because of water being impounded to serve Florida's tremendous growth, but the salinity of this estuary has been changed for the worse. To the north, the engineers are engaged in an outrageous piece of pork barrel that is flimsily justified on the grounds of recreation. They are more than one fifth of the way done in cutting a 107 mile canal from Yankeetown on the Gulf of Mexico to Jacksonville on the Atlantic. Called the Cross-Florida Barge Canal, this $177 million boondoggle appears likely to drastically alter two river systems—the Oklawaha inland and the St. Johns that empties into the Atlantic as the source of a major shrimp industry.

The FWQA Report lists twenty-five estuarine systems as having *definitely* been degraded by pollution. It also notes that the remaining 38 percent of U.S. estuaries may well be damaged ecologically, but "there is just no easily identifiable pollution problem present." Here are the twenty-five definitely degraded estuaries, making up 62 percent of our coastal waters:

Penobscot Bay	James River
Boston Harbor	Charleston Harbor

Moriches Bay   Savannah River
New York Harbor   Biscayne Bay
Raritan Bay   San Juan Harbor, Puerto Rico
Delaware Estuary   Tampa Bay
Baltimore Harbor   Pensacola Bay
Potomac River   Mississippi River
Galveston Bay   Columbia River
Laguna Madre   Puget Sound
San Diego Bay   Silver Bay, Alaska
Los Angeles Harbor   Hilo Harbor, Hawaii
San Francisco Bay

Runoff, Wastes, Thermal Effects, and Catastrophes at Sea

Rivers continue to be the handiest sewage and waste outlets. It has always been assumed that once downstream and out to sea, the filth would dissipate or be broken down biologically over a period of time, since the volume of the ocean was so immense. The FWQA report notes matter of factly that "Virtually all of the cities and industries in the coastal counties dispose of wastes either directly or indirectly into the estuarine zone." But as coastal fisheries have declined and evidence of serious contamination has been found offshore and in the estuaries, officials have become alarmed. The U.S. Coast and Geodetic Survey recently discovered that the Gulf Stream was farther away from Miami than the city had assumed when counting on the current to carry away all of its sewage and municipal wastes. Miami's beach seekers could thus be exposed continually to hepatitis from live viruses caught inshore. New York's lethal sludge was cited in Chapter 5, and there are countless such instances. The Chesapeake oysters and crabs and the Gulf shrimp and other coastal fisheries are directly threatened by pollution.

Insecticides and other chemicals run off farmland or insect control areas into rivers and streams and thence to lakes and the ocean. That there was any danger in this was vainly pointed out by the late Rachel Carson nearly a decade ago and made dramatically apparent when the federal government seized a huge shipment of Lake Michigan coho salmon, contaminated by DDT, in the spring of 1969. In the past year, new findings have confirmed the worst fears of Rachel Carson and spell trouble for the estuaries. Nitrates, phosphates, and other minerals naturally *contribute to estuarine fertility only in very definite amounts,* depending on the water characteristics and movement of the particular area. As noted in Chapter 1, the Great Lakes have deteriorated because of *too much phosphorus.*

A major threat on the horizon is thermal pollution, defined in Chapter 1 and expected to magnify considerably as some forty-seven nuclear power plants are now being built or planned for completion by 1976, producing a total of thirty-five thousand megawatts of power. Most of these are being contested. For example, conservationists have organized to fight the Baltimore Gas & Electric Company's plan to build a nuclear plant at Calvert Cliffs, Maryland (one of fifteen planned for the Chesapeake Bay). While the state was satisfied that this facility would not pose a thermal hazard after the electric company had added some environmental precautions to its original proposal, the Calvert Cliffs plant is still the subject of a court suit, as will be noted in Chapter 10. The citizens, including the President of the United States, who have houses on Biscayne Bay are in an even more serious predicament. Already, two fossil-fuel (conventional) power plants on Turkey Point have destroyed fish, mollusks, and lobsters near the point where they discharge hot "cooling" water into the bay. Federal Water Quality officials became most alarmed, then, when Florida Power & Light Company began to build two nuclear generators at Turkey Point.

According to FWQA biologists, thermal discharges from these plants of up to ninety-five degrees would kill fish life throughout the bay. The Interior Department has brought action against the utility, seeking to ensure that heat discharges be kept to within four degrees of the bay fall, winter, and spring temperature, one and a half degrees in summer when the bay temperature is in the mid-eighties.

Oil

Despite previous assurances that such events were rare and unlikely, within a year there were two major oil spills from drilling rigs on the Continental Shelf off the U.S. The Santa Barbara eruption in the winter of 1969 was a shocker. It created such a mess, such an outcry, that a year later the President announced that twenty drilling leases would be cancelled in the Santa Barbara Channel so that an area could be set aside as a marine sanctuary in addition to zones previously set aside. Nixon's measure unfortunately did not limit drilling in fifty other channel leases or place restrictions on a large offshore area where the federal government has yet to sell oil rights. The second disaster occurred on a Chevron Oil Company rig off Louisiana's rich estuaries. The federal government has sued the oil company for failing to use a safety valve called a "storm choke," that would have prevented the accident on the runaway well and was missing on 136 others in the area.

Ever since the Torrey Canyon broke up off Cornwall, England, and spilled oil that killed thousands of birds and destroyed shellfish along the French coast, there have been quite a number of conferences to devise techniques as well as laws to cope with a serious threat to the world's oceans. And throughout the aftermath of both the Santa Barbara and Gulf catastrophes, commissions, legislators, and scientific bodies have attacked the problem in the U.S. Nothing has resulted that significantly would reduce or prevent oil spillage. In other words, there have been no strictures to drill for oil only on land, and oil facilities and transport methods remain basically unchanged.

However, as a result of pressure from conservationists, the 1970 Water Quality Improvement Act contained several provisions concerning oil damage. Ship owners or companies handling oil at onshore and offshore facilities bear absolute liability in case of spillage, unless someone else is to blame, or it is an act of war or God. Cleanup reimbursement runs up to $14 million (or $100 a gross ton—whichever is less) in the case of a ship, and up to eight million dollars in the case of a facility. In addition, the act required that the President publish a National Contingency Plan providing the Coast Guard with a $35 million revolving contingency fund and a trained team of specialists to go into action immediately in the case of spill.

A few accidents widely spaced, as to time and place, are unfortunate but they do not present a long-term threat. What alarms biologists is that the number has increased tremendously and the situation can only get worse with the advent of supertankers twice as big as conventional carriers. It is estimated alone that ships discharge some 400,000 tons of oil a year by illegally pumping their bilges or by accidentally opening a hose or valve.

One of the most comprehensive but succinctly stated summations of the effects of oil on the marine environment was prepared for the Canadian government by Dr. Richard E. Warner of the University of Newfoundland. (See Bibliography.) He specifically was concerned with the effects of oil in the cold water estuaries of the northern hemisphere and in the frigid Arctic. But he cited case after case of oil operations adversely affecting commercial fisheries, waterfowl, and seabirds in warmer waters where oil biochemically decomposes and disperses much more quickly. Then he noted that the decomposition rates slow down at lower temperatures, practically coming to a halt at freezing. "Over half the earth's surface has mean temperatures well below those necessary for the optimal biochemical decay rate of crude oil," wrote Warner. "And decomposition in the Arctic oceans, whose temperatures are at 0°C (32°F) or below throughout the year, would be very slow indeed." In addition to oil in Arctic waters having persistent effect, Warner noted that in the cold climate the more toxic lighter hydrocarbons in the oil would not evaporate quickly.

In Alaska's Cook Inlet, U.S. fishermen have already suffered losses of fish and crab. Thousands of seabirds have died, and biologists are worried about the effects of one or two small spills a fortnight on beluga whale, seals, sea otters, bears, and other fur animals that live in the coastal environment. The danger to Alaskan estuaries from oil spill will increase as more oil is taken from the North Slope. Warner and his colleagues shudder at the thought of a tanker breaking up in Arctic waters while using the Northwest Passage. They also fear that landborne pipelines and onshore facilities could be easily ruptured by the ice and terrific ground contractions and heaves that are common in the north. If this happened, as the U.S. Geological Survey has warned, the damage to hundreds of river estuaries, marshes, and wetlands would be incalculable. The Alaskan north is the summer ground for millions of migratory waterfowl whose population might be seriously reduced by an oil catastrophe. Some seventy thousand greater snow geese, most of the world's remaining population, just missed such a fate in August, 1963, when a ship illegally dumped only one thousand gallons of fuel into the St. Lawrence River in Canada. Through a massive effort, the oil was cleaned from the birds' nesting area only a few days before they arrived.

WHAT IS BEING DONE?

In spite of all the documentary evidence of damage already done and likely to occur in the nation's marshes, wetlands, and estuaries, surprisingly little has been instituted at any government level to cope with the problem. Under the Water Quality Act of 1965 the federal government can only intervene in coastal waters or in the Great Lakes when an interstate waterway is violated, when interstate commerce is affected, or when a governor issues an invitation. The second reason applied to the polluted shellfish in the Gulf of Mexico. The Interior Department was able to call an enforcement conference in Texas on the grounds that economic injury resulted because shellfish could not be marketed in interstate commerce. The states have brazenly polluted their shorelines by tolerating industrial contamination of seaward flowing rivers and by failing to urge land use controls on the waterfront. Pending in Congress are proposals for national water quality standards that would enable the federal government to crack down on intrastate pollution, coastal and otherwise.

While the Interior Department can acquire shoreline property for parks, seashores, or wilderness areas, *land use* in the estuarine areas is up to the local governments and states. But *landfill* and dredging activities present another opportunity for action. This is made most clear in a report issued by the House Committee on Government Operations (see details in Appendix I) explaining how the Army Corps of Engineers could be a constructive force in preventing pollution of coastal waters and wetlands.

Rivers, Harbors, and Refuse

Under the 1899 River and Harbor Act, the Corps of Engineers had the power to regulate any filling or dredging activity right up to the high-tide mark and could act to prevent the discharge into a navigable waterway "any refuse matter of any kind or description whatever other than that flowing from streets and sewers and passing therefrom in a liquid state." (This refuse clause was also discussed in Chapter 1.) Over the years, the Corps narrowly interpreted the Act as allowing any filling or dredging permit provided it did not interfere with or obstruct navigation. The engineers have persisted in overlooking environmental side effects and have failed to do anything about the "refuse matter" section of the law right up until 1969, despite a court decision in 1933 that defended public and recreation interests and despite a spate of memoranda between the Interior Department and the Corps agreeing that the environment must be considered when granting permits. Now, as a result of the 1969 National Environmental Policy Act, the Corps has been more cooperative. The key section of the Act is 102, which is a mandate to all federal agencies to give consideration to environmental values in taking any action and to provide a "detailed statement" along with every recommendation that might affect the environment. This statement must judge recommendations in accordance with resource and ecological criteria and their effects on human health and productivity.

As a result of the House Committee report, the Corps agreed to amend its regulations regarding applications for sewer pipelines that discharge into navigable waterways. Henceforth the effluent must be fully described so the Corps can judge the sewage in terms of water quality criteria. The eighteen page document is a pamphlet well worth obtaining for inclusion in one's antipollution kit. It noted, for example, that the Corps has for years also established harbor lines in estuaries for no other reason than to fix a boundary beyond which no bulkheads or piers could be built. Yet in every "harbor," there are miles of wetlands or marshes behind these boundaries to which the Corps has paid no attention. The report, which was prepared at the request of Henry Reuss, the crusader cited so many times in this book, made eight recommendations: that the Corps should (1) generally learn to consider ecological values in granting dredging and landfill permits; (2) permit no further such operations in the nation's estuaries unless an applicant proves his project is in the public's interest; (3) revise the harbor lines definition so that applications for work behind the boundaries will be judged by appropriate standards; (4) grant public hearings concerning harbor line changes; (5) deny permits for sewer outfalls unless water quality criteria are to be maintained in the waste content; (6) enforce the Refuse Act of 1899; (7) request the attorney general sue those who do not promptly comply with the Refuse clause; and (8) increase its capability for cleaning up estuarine pollution. If the Corps of Engineers were to follow these, the destruction of thousands of acres of marshes, tidal lagoons, sandbars, and shallow bays by filling and dredging would *cease*. Then maybe the Corps would have time to join the battle against the other kinds of pollution by helping to construct sewer systems and treatment facilities along the coast, along rivers that empty into the sea, and in the Great lakes. The engineers have mastered the government's bureaucracy. They know all the ins and outs of getting requests through Congress and the money to implement their plans. This is evidenced by a Public Works budget of (nearly five billion dollars) four times the amount being spent annually to combat water pollution.

More Studies and Plans

For over a decade, ecologists, marine biologists, oceanographers, and conservationists have written and testified about the crisis along the shoreline. Organizations like Ducks Unlimited and Nature Conservancy have quietly purchased marshes and wetlands to preserve important wildlife habitats. In cooperation with the Canadian government, Ducks Unlimited has acquired over two million acres of prairie marshes and lowlands, mostly in the middle provinces just north of the U.S. border, and to a lesser degree in the maritime provinces and the far western British Columbia. The organization has put over $14 million into small holding dams, dikes, and trenches to convert this land into wetland and marsh breeding spots, called "duck factories," so that the waterfowl could survive and fly south to U.S. estuaries and lakes. By default of government action to protect a public interest, private groups have performed a valuable service in saving coastal lands. A good example of this approach is the plan of the Little Cumberland Island Association, which has preserved an island off the Georgia Coast that is one of the few remaining beach sites where loggerhead sea turtles lay their eggs and which offers fertile marshes and ponds to shore birds and waterfowl. The group consists of shareholders who can build a cottage or obtain lodging on the 10 percent of the island where development has been allowed and carefully planned. "The objective here is to maintain a private, coastal, wildlife preserve with the support of a limited number of people who are thereby privileged to enjoy living on the edge of it without disturbing the natural ecology of the area," says Ingram H. Richardson, president of the association.

The touch-and-go plight of the whooping crane and the

persecution of the alligator have dramatized the uniqueness and value of estuarine habitats. And the great surge to the waterfront during comparatively prosperous times and in an increasingly mobile and recreation conscious America ought to have alerted the public to the threatened coastal environment. Lake Erie was the first of the Great Lakes to become foully polluted. Lake Michigan is slowly being degraded and even Lake Superior, the clearest and biggest, now is tainted.

Yet in spite of these events, and any number of federally initiated studies and task forces examining the problem, the only legislation from Washington has been a law directing the Department of Interior to conduct still more studies and come up with proposals regarding estuarine protection. The law's title, Estuary Protection Act of 1968, is thus misleading, although Congressman John Dingell and others pushed hard and in vain to give this measure some bite. As a result of knowledge assembled by the FWQA and the Fish and Wildlife Service, the Nixon administration has proposed new legislation that prompted Dingell to remark, "The Administration has failed to come up with a program." The proposal would provide the coastal states with planning grants to devise comprehensive shoreline zoning and long-term land use plans. However, the grants would be offered only to a state that had already begun its own estuarine program satisfying federal environmental criteria. Furthermore, while states would have to comply with environmental criteria in order to receive the study grants, it would not be mandatory for them to get together with the federal government at all to protect the coastal zone. Other proposed bills do not go much further. So, if no stronger legislation is proposed, these measures ought to be encouraged and hopefully strengthened, but in sum, there is nothing pending to force states to take action to protect their shoreline wetlands. Excellent summaries of what has been discussed and what is now proposed in Congress are provided by the Conservation Foundation and the League of Women Voters. (see Bibliography.)

In the process of pushing government reorganization, the Nixon administration has called for both an Environmental Protection Agency and a National Oceanic and Atmospheric agency. Possibly the latter may move aggressively to deal with the shoreline. One of many studies, "Our Nation and the Sea," by the Commission on Marine Science, Engineering and Resources, set up by the Johnson administration, recommended this department. It proposed a "coastal management act" to provide leadership and aid to state coastal protection and an agency with administrative and funding powers. This study put the entire management problem succinctly when it noted that "The rapidly intensifying use of coastal areas already has outrun the capabilities of local governments to plan their orderly development and to resolve conflicts. The division of responsibilities among the several levels of government is unclear and the knowledge and procedures for formulating sound decisions are lacking."

The FWQA study, cited previously, was ambiguous. It concluded that the federal role was "not the primary one" in protecting the estuaries. But it also painted a picture which leads to the conclusion that, while state governments should be given a free hand to take the initiative, the federal government must be able to wield a big strong stick. The reason, as noted in the report, is that "The economic, social, and environmental use and well-being of the estuarine and coastal zones of the nation are of vital interest to the *inland states as well*." (Emphasis added.)

In other words, since the shoreline is a *national boundary, a national resource,* there has got to be a focal point of responsibility and *authority*. The Interior Department's recommendations are commendable, but unless states and local governments are levered into compliance, no federal program can be very effective.

Two Great Bays

The experiences of two estuarine regions should *not* have to be repeated, even though in one case citizens fought day and night to achieve what appears to be a splendid victory. This was on San Francisco Bay. Hanging in the balance, with not much hope as of the moment, is the fate of the nation's most diverse estuary, the Chesapeake Bay.

People come to San Francisco to gaze out at the wind-whipped currents that swirl under the Golden Gate, past Alcatraz and Angel Islands to meet the fresh flow of the Sacramento River. In the past, but no longer, the bay was rich in crabs and shrimp. But in 120 years since California statehood, dikes and landfill projects have reduced the bay's area from 680 to 430 square miles. As noted already, 192,000 acres of wetland were destroyed between 1950 and 1968 out of 294,000 acres considered to be valuable shoreline habitat. Industries, residential developments, airport runways, and urban expansion gobbled up great hunks of the bay. Since 70 percent of this estuary is less than eighteen feet deep, it was comparatively inexpensive to "reclaim" for waterfront real estate. And, as Congressman Paul N. McCloskey, who wrote the preface of this book, and state legislators can readily attest, the pressures on communities to develop the bay are compelling and hard to stop. With a rapidly growing population (from 2.7 million to five million between 1950 and 1970 and a projected population of 7.5 million by 1990), some 36 communities felt they had to have the tax revenues generated by new development, particularly valuable waterfront property, and construction.

A truly remarkable outpouring of conservationist sentiment as well as working alliances between many groups, leadership by two legislators, the late Eugene McAteer and Nicholas Petris, and the support of other officials led in 1967 to a reprieve. A bill sponsored by McAteer and Petris was passed, creating a commission to rule on bay-fill proposals—temporarily setting a moratorium on such projects—while recommending permanent legislation to manage and protect the bay forever. The so-called Bay Conservation and Development Commission expired in 1969, but its proposals were accepted in 1970 after a last ditch battle against development interests.

The BCDC then was established as a permanent management body ruling on development of most wetlands and the shoreline 100 acres landward. Except for some urban expansion, new airport runways, and the reservation of land for industrial development, the bay should avoid the fate of being filled in completely but for a few navigation lanes, the gloomy prospect not long ago. The agency has worked out a master plan for bay protection and development which is, to be sure, an accommodation of most interests and will compensate property owners for land eventually acquired for public enjoyment.

Nothing like this has yet been generated on the Chesapeake where planning obstacles are compounded because the shoreline is shared by two states, Maryland and Virginia. The bay geologically is the drowned mouth of the Susquehanna River that now empties at Havre de Grace, 165 miles up the estuary from where the bay meets the Atlantic near Norfolk. The Chesapeake has 4,500 miles of shoreline because of its many inlets and myriad number of rivers, including such large waterways as the Patuxent, Potomac, Rappahannock, York, and James. It averages less than twenty feet in depth and has thus been as attractive as San Francisco Bay to speculators.

During recent decades the pressures of some ten million people have visibly deteriorated both the water and land environment of the estuary. Farmers on the eastern shore peninsula —which has been connected at opposite ends by bridges—are being taxed out of business and lured by developers to sell off marsh and wetland waterfront. Coves and creeks have become over enriched from farm and sewage runoff or filled in with silt. Oysters, clams, and crabs have declined markedly. This past summer the death of some thirty thousand crabs in Nassawakox Creek on the Virginia eastern shore was blamed on DDT. Oyster beds are condemned frequently because of pollution and fish kills are just as common. There are at least nine marine laboratories or research stations on the bay, because its ecology is so diverse and fruitful, but so far the warnings of the scientists have not been translated into action. Instead, as already noted, the bay is threatened by more power plants which could warm up the water enough to destroy the most important marine life. Domestic and boat sewage has loaded the bay with bacteria and nutrients. Baltimore industries add chemical wastes and a heavy volume of shipping regularly deposits oil and other substances, the results of illegal bilge pumping and general carelessness.

Tentative progress has been made. Maryland has passed compromise measures to protect wetlands and regulate dredging and landfill projects but Virginia has done virtually nothing. Both states have different conservation laws. For example, dredging for crabs and purse seining for menhaden is forbidden in Maryland waters during the summer, but fishermen simply cross the imaginary line drawn across the bay and pursue these activities legally in Virginia. It is only too clear that a Bay Commission or Chesapeake Basin Authority is needed to proscribe conservation rules and protect the marshes and wetlands in this estuary discovered by Capt. John Smith.

What Have the States Done?

Even though local interests and pressures are intense, coastal states have passed laws or conservation regulations that directly or indirectly affect the health of their shoreline environment. They are Massachusetts, Maine, New Hampshire, Rhode Island, Connecticut, New York, New Jersey, Maryland, North Carolina, South Carolina, Florida, Alabama, Louisiana, Texas, California, Wisconsin, Oregon, Washington, and Georgia.

The first and the strongest legislation was passed in Massachusetts in 1963. It was a coastal zoning law that allowed the state department of natural resources to review all dredging, filling, or alteration of the shoreline. The act was amended in 1965 to give property owners adequate compensation rights and to give the state authority to actually *take the initiative* in coastal planning by issuing specifications on land use and restricting wetland development before it was contested.

The conservation commission movement, cited in Chapter 7, has been a powerful force in obtaining laws to protect the coastal strip. In Connecticut, dredgers of sand and gravel for construction must apply for a permit and may be required to post a performance bond by the Commissioner of Agriculture and Natural Resources, who judges the application in terms of its ecological effects. Rhode Island has passed an Intertidal Salt Marsh Law prohibiting disturbance of marsh ecosystems by dumping or excavating. Maine this year passed a law providing strict controls on the development of the coastline by oil and petrochemical industries. New Jersey has declared a moratorium on wetland development, pending completion of a study of all state-owned tidelands.

An extraordinary ruling was handed down in Georgia in March, 1970 by state Attorney General Arthur K. Bolton. He declared that the state held the coastal areas in common-law trust for the public and thus could restrict any kind of development of marshes and wetlands. He based his position on the English crown's control of coastal property for public benefit.

Already in Oregon, the state supreme court has ruled that the public has a right to all coastland up to the vegetation line.

Previously, the public had held beach rights below the tide mark.

WHAT CAN YOU DO?

These are samples of approaches that have been taken, but they are by no means the ultimate solution. Nothing short of a national shoreline master plan with *controls—not just incentives* —to ensure state cooperation will really be effective. This is an objective that you can discuss or support in a letter to your congressman, senators, and the White House environmental Quality Council. You can of course also push local and state measures similar to those passed in the states just mentioned.

Within the present system, there are a number of recourses. If a federal agency is involved in an estuarine project, directly or through aid and state permission, you can cite Section 102 of the National Environmental Quality Act, which, as noted, requires a statement of environmental responsibility. Under the provisions referred to in the Water Quality Act, you can also seek federal intervention. Land use laws are just as important, and you can ask your local planning officials what shoreline is accessible to the public now, and potentially, through easements or acquisition. The example of the conservation commissions in New England is to be emulated, as I have emphasized in Chapter 7.

If the nation's founders had been able to assess or grasp the natural qualities of this land and the encroachments to be faced, and if they had been in any position to dictate the patterns of land development, the shoreline today would be in a far different state. Probably it would have been developed exclusively in accordance with the English doctrine of Public Trust. Private ownership would no doubt have been forced to follow strict zoning rules, keeping the beaches and marshes open except for necessary piers to the ocean. Deep water harbors would have been located to provide the most logical displacement of industries, cities, and transportation terminals. Long open beaches would have been kept free of concession stands and strip motels but open to the public. Weekend, vacation developments would have been clustered back far enough to allow the shoreline to provide shelter for fish and wildlife—without the effects of pollution—and a scenic retreat for human beings.

A shore and marsh ethic is long overdue. Already too much of the U.S. is bordered by bulkheads, rows of housing, ugly factories and power plants, and garbage dumps fit only for rats and seagulls.

9 ❧ FARMING AND GARDENING FOR SAFETY

THE USE OF SYNTHETIC pesticides began as a miracle. It has steadily become a nightmare. DDT, the forerunner, was given credit for saving thousands of lives by combating typhus, malaria, cholera, Rocky Mountain spotted fever, encephalitis, and a host of other diseases. It was massively, indiscriminately applied to crops to save food for starving, overpopulated countries. But then the old disease carriers and pests became immune to the treatment. Where they once had worked, DDT and other synthetics lost their touch. Insects previously controlled by natural balance became new pests as their predators were eliminated. Alarming side-effects on birds, fish, and other animals were reported, and now *human beings* appear endangered.

It is a mean twist that, like drugs, pesticides have created a dependency. While the habit can certainly be kicked in the home garden, some synthetic substitutes for DDT will probably have to be applied at a reduced rate, selectively or in combination with natural controls, on crops and *maybe* in the epidemic-prone areas of underdeveloped countries. Because while we've been on the pesticide treadmill, the disease-bearing, crop-destroying insect populations have grown, and their natural enemies have decreased.

Even then, the pesticide problem is fraught with uncertainty. New villains appear as fast as the old ones are diagnosed. As the Nixon administration's commission on pesticides concluded in its summary report (see Bibliography), "The field of pesticide toxicology exemplifies the absurdity of a situation in which 200 million Americans are undergoing lifelong exposure, yet our knowledge of what is happening to them is at best fragmentary and for the most part indirect and inferential."

The pesticide story is a sad parable, a pessimistic commentary on our environmental sense of responsibility. *Seven years* passed between the documented warnings of the late Rachel Carson in her book *Silent Spring* and the explosion of widespread public concern over the ecological effects of pesticides such as DDT. Scientists and legislators such as Wisconsin Senator Gaylord Nelson had to beat the drum thin to gain vindication for the much maligned woman biologist. If it takes that long for the environmental issues of 1970 to be translated into forceful action and thoroughly ingrained in the public consciousness, then the political and social upheaval that also besets America may be irrelevant. Each of the environmental problems described in previous chapters likely will have escalated beyond control from the effects of constant, dulling debate and inaction. Sadder yet, even our attempts to deal with the pesticide crisis so far have fallen short of what the great majority of ecologists feel is necessary to avert more silent springs in future years.

GENERAL EFFECTS OF PESTICIDES

So much has been written about DDT and other "persistent" pesticides during the past two years that it would not be productive in this book to present more than a summary of what these chemicals do. The politics of pesticides is another matter which will be taken up here and in Chapter 10 on The Law. Charles F. Wurster, chairman of the scientists' advisory committee of the Environmental Defense Fund and a professor of biological sciences, summed up the feelings of many who have led the long fight against DDT in an article in the May 4, 1969, *Washington Post*. He noted that since 1946, biologists had warned that common spraying programs would kill birds. By now, Wurster wrote, "The process has been so thoroughly studied and documented by so many scientists in so many parts of the country that it is no longer of scientific interest."

DDT is toxic, persistent, mobile, and soluble, Wurster noted in that same article. It poisons all animal life to some degree and it does not break down biochemically for a decade or more. It is carried long distances in both the water and air and it is easily absorbed and then retained by living organisms. However, DDT does *not* dissolve in water, a liability that is shared by other chlorinated hydrocarbons—dieldrin, aldrin, endrin, heptachlor, chlordane, and lindane. They of course are just as dangerous.

Silent Spring and other texts noted in the Bibliography are full of examples of what DDT and other chlorinated hydrocarbons (or organochlorines) have done to the creatures of our environment. From these accounts and my own research, here are a few typical cases:

≈ British biologists in the early 1950's linked DDT and broken eggs found in the nests of peregrine falcons and sparrow hawks, which were naturally declining because of this odd mishap. Since then, further research has shown that DDT causes an enzymatic reaction that inhibits the calcium production of the bird and thus results in a thin-shelled egg, or a shell-less membrane as was found in an eagle's nest in northern Michigan in the spring of 1969. Falcons, certain hawks, and the bald eagle—so-called birds of prey—feed on small living creatures that have absorbed DDT from tiny organisms. The amounts accumulate successively upward through the food chain, magnifying to a considerable degree in the final predator. The California brown penguin and the Bermuda petrel as well as the falcon and the bald eagle are becoming rare species because of DDT.

⅜ Wurster and fellow biologists at the State University of New York and Brookhaven National Laboratory traced a graphic example of "biological magnification in an estuary of Long Island Sound. Plankton contained 0.04 parts per million of DDT; small predator fish, 0.23 ppm; larger fish 2 ppm and ospreys at the end of the chain, 13.8 ppm.

⅜ Spreading the affliction is one thing. Aiding pests by killing natural predators is another. It boomerangs directly on the person or government that uses a persistent pesticide. Thousands of acres of forest were cleared in the mid 1950's to start growing cocoa on a grand scale in Malaysia. The brushy undergrowth that preceded the return of the taller trees around the cocoa fields harbored pests that attacked the new crop. A DDT formula was sprayed, with initial success. But then the pests struck back and over a three year period did more and more damage despite repeated sprayings. What had happened? The contact-acting spray had killed as many predators as pests, while through the evolutionary selection process a breed of super-resistant pest developed. As the forest growth took over, it should have been a refuge for predators to destroy the pests in the underbrush. But the spraying had killed off the predators. Eventually all but one plantation dropped DDT. That one followed suit after the pests had become even more rampant during more massive applications of DDT.

⅜ Similarly, the use of chlorinated hydrocarbons backfired in the Canete Valley in Peru. Within five years, the chemical had increased the cotton yield 50 percent above average to a high of 649 pounds per acre in 1954. But suddenly and strangely, the number of pest species doubled and the old pests were resistant. In 1955, the yield dropped to 296 pounds an acre and repeated pesticide dosages failed. Biological controls coupled with a dramatic reduction in the use of DDT brought the cotton crop back to record production, 923 pounds an acre, by 1960.

✻   In U.S. cotton fields, DDT has been the main weapon against the boll weevil, but here too unexpected things have happened. The weevil seems to have developed a resistance to chlorinated hydrocarbons, and two species of bollworms that previously caused little concern have become serious pests. A Shell Chemical persistent pesticide, Azodrin, was touted as being the answer to the problem of lygus bugs in California's San Joaquin Valley. But according to a report by Stanford biologists Donald Kennedy and John Hessel in *Cry California* (see Bibliography) the lygus bug has been found *not* to significantly decrease cotton yields while the azodrin likely killed natural pest predators such as green lacewings, lady bugs, and pirate bugs, which feed on the eggs and larvae of the pink bollworm. Agricultural experts have conceded that the pesticides' role in the war against the boll weevil will have to be limited.

✻   The appearance of unusual levels of DDT in city soil (e.g., Cincinnati) and in the tissues of remote creatures such as the penguin in the Antarctic and the polar bear in the Arctic, have convinced scientists that the chemical is most dangerous and unpredictable as an air pollutant. Evaporated as a dust particle, it returns to the earth with rain and snow. However, it has produced the most spectacular effects in the aquatic environment. In the winter of 1969, the Food and Drug Administration, which had for years turned a deaf ear to criticism of DDT, confiscated twenty-eight thousand pounds of coho salmon caught in Lake Michigan and contaminated with levels of DDT well in excess of tolerances set for meat and other products. About the same time, reports appeared noting that DDT had wiped out lake trout fry in Lake George, N.Y., and killed up to a million salmon fry in Lake Michigan by poisoning the yolk sac, the newborn fishes' only sustenance in their first days of existence. Since then, Department of Interior investigators have announced that pesticides used to control fire ants in southern states were lethal to blue crabs and shrimp, the South's most important fish resources. The Bureau of Commercial Fisheries has found that minute amounts of DDT stop oysters from feeding and inhibit new shell formation. Most alarming of all, however, has been the discovery that chlorinated hydrocarbons upset the photosynthesis of marine phytoplankton, which are essential tiny organisms that begin the oceanic food chain and produce its oxygen.

MAN, THE ENDANGERED SPECIES

This was the theme of the Interior Department's 1968 yearbook. As each day passes, it seems less extreme as a prophecy for the future. While the content of persistent pesticides is fractional in fish and wildlife, it is believed to average ten to twelve parts per million in the fatty tissues of each American. There is as yet no clinical proof that measurable adverse health effects are the result and, needless to say, human beings are not used experimentally to find out. However, tests with animals have proved most upsetting.

⚜ Testifying before the Wisconsin Department of Natural Resources during 1968 hearings as to whether DDT should be banned in that state, Dr. Robert Risebrough of the University of California's Marine Resources Institute, described the adverse influence of pesticides in inducing enzymatic reactions in the liver, thus affecting sex hormones (estrogen, progesterone, and testosterone) in unknown ways. "No responsible person could now get up here and say that this constant nibbling away at our steroids is without any physiological effect," Risebrough said. "It would be irresponsible."

❧ In a petition to the Food and Drug Administration in October, 1969, requesting that the government enforce the Food Additives Clause (Delaney Amendment) by setting a zero tolerance for DDT in foods, the Environmental Defense Fund cited a number of studies showing that the pesticide was carcinogenic in rats, mice, and trout. Moreover, the petition noted, studies at the University of Miami School of Medicine revealed that human victims of terminal cancer contained up to twice the presumed national average of DDT in their fat. It is hardly comforting that the medical school did not establish a direct correlation between the pesticide and cancer.

The Nixon pesticides panel was impressed by such evidence and commented that "After carefully reviewing all available information, the commission has concluded that there is adequate evidence concerning potential hazards to our environment and to man's health to require corrective action. Our nation cannot afford to wait until the last piece of evidence has been submitted on the many issues related to pesticide usage." Unfortunately, ensuing government action was not only ambiguous, it allowed yet another long wait before "corrective action" could be enforced against irresponsible use of DDT.

ECONOMIC AND POLITICAL REALITIES

While at least one maker of DDT, Olin Corp. (twenty-five to thirty million pounds, or 20 percent of U.S. annual production) has decided to drop the chemical, ironically because of adverse ecological effects discovered near its DDT plant in Huntsville, Ala., the production and use of chlorinated hydrocarbons is still big business. Worse, they are exported to underdeveloped countries where U.S. advisors tout their benefits, but give little or no warning at all about potential hazards. And of course persistent pesticides are not the only undesirables. Short-lived substitutes and other varieties have equally devastating effects, as will be noted further in this chapter. It was estimated that U.S. farmers would spend about $1.5 billion on pesticides in 1969, more than double 1968. About a quarter of the volume produced were chlorinated hydrocarbons and about one fifth of the total tonnage was used in the household and garden—the rest being applied by agriculture (three fifths) and industries and institutions. Pesticide proponents point out that for every dollar spent on chemical protection, up to five are saved in increased yields. This and other extravagant claims about the value of pesticides in the so-called "Green Revolution" *do not take into account* the costs of adverse side-effects that result from occupational accidents, immediate damage from drifting or seeping chemicals, and long-range accumulation in wildlife and humans. The economics of agriculture are always uncertain, and competition to serve this industry is intense. For these and other reasons, the profit margins in sales of pesticides for home application are the highest. This is a market in which, regrettably, garden chemical manufacturers are most easily able to take advantage of consumer gullibility and ignorance with misleading advertising pitches and with a mind-boggling, ever-changing array of trade names.

They get away with it because pesticide regulation is truly a joke. A structure exists to control agricultural use, even if it has not yet been vigorously implemented. But the poor home-owner is invariably taken for a ride. The Food and Drug Administration of the Department of Health, Education and Welfare has authority to set pesticide tolerances in food; the Environmental Health Services of HEW advises on drinking water, and the Interior Department *advises* concerning effects on fish and wildlife. *However,* none of the information provided by these agencies is reflected in labeling instructions or required in advertising the chemical cure-alls.

The chief administrator of pesticides policy has in fact been the Department of Agriculture even though this agency's mandate is to increase farm production and uphold the economic interests of farmers. Agriculture's Pesticides Regulation Division is supposed to carry out the provisions of the Federal Insecticide, Fungacide and Rodenticide Act (FIFRA), which actually does no more than require pesticide manufacturers to follow labeling and registration rules. A label is supposed to tell you how, on what, and when you use a chemical, with warnings as to its toxicity. As noted, a label does not, for example, tell you that what you're buying is a potential threat to water supplies or your food, if it is used indiscriminately.

Actually, the Department of Agriculture *can* cite FIFRA to keep dangerous pesticides off the market. But is never has, until after HEW has discovered a case of food contamination. The General Accounting Office issued a scathing report on the DA's handling of pesticides in the fall of 1968, noting among other things that the agency failed to maintain an adequate inspection and follow-up system and that in thirteen years not one violator of FIFRA had been prosecuted, *despite repeated violations* and scores of citations.

Yet as a result of high-level concern during the past year and a half, the public is now under the impression that the system is changed and that the pesticide problem is under control. Such is far from being the case, although recent court decisions (see Chapter 10) promise relief. Briefly, here is the chronology of inaction and political indecision.

As a result of the pesticide panel recommendations, in November, 1969, stronger interagency cooperation was pledged, and by agreement HEW was accorded new power in determining the requirement of health safety. More importantly, the administration ordered that the "use of DDT and DDD (a derivative) be restricted within two years to those uses essential to the preservation of human health or welfare." However, "essential" was a term never defined and a good deal of latitude was provided to meet this objective.

This was apparent when in response to a petition calling for immediate suspension of DDT registration, DA published a notice in the *Federal Register* saying that it was *"considering" cancellation* of all but "essential" uses of DDT for health safety subject to comment by "interested persons" within ninety days. This gave manufacturers and users time to muster their arguments. Moreover, a long time-lapse could be expected during which comments were weighed. Cancellation may sound more final than suspension, but it is not. It allows those who object to the order to file court appeals and buy more time. (One manufacturer's appeal is now pending.) On the other hand, a suspension would have amounted to an immediate ban until DDT was proved safe beyond a shadow of a doubt.

2, 4, 5-T AND 2, 4-D

So DDT products remain on the shelves, mocking the cancellation order, but even if DDT was suspended, the government does not have the inspection force necessary to confiscate these vast and scattered stocks. Similar action against the herbicide 2,4,5-T has suffered the same fate. The government last fall first became concerned about this weed killer used as a defoliant in Southeast Asia, after it was found to cause birth defects in animals. As this evidence was reinforced (a first-rate series of disclosures by Thomas Whiteside appeared in *The New Yorker* magazine), in April, 1970, the Department of Agriculture and HEW jointly declared *suspension* of *liquid* formulations of 2,4,5-T used around the home and in aquatic ecosystems (around the edges of ponds and along ditches and canals). *Powder* formulations and mixes in fertilizers were allowed to continue in use in lawns, gardens, and fields, provided water supplies were not endangered. Then at a Senate hearing the U.S. Surgeon General, Dr. Jesse L. Steinfeld, said that DA would cancel nonliquid formulations of 2,4,5-T used around the home and on food crops. Finally the Defense Department announced it had stopped using 2,4,5-T in Southeast Asia, where already the defoliant had wrought ecological havoc over millions of acres of jungle and rich productive countryside.

The Pentagon said, however, that 2,4,-D, a member of the same family of polychlorinated phenolic compounds considered less harmful, would continue to be sprayed to defoliate sparsely populated areas.

What has happened since these pronouncements? Ralph Nader's Center for Responsive Law has filed a petition against the Department of Agriculture, contending that the ban on 2,4,5-T neither went far enough nor had been enforced. The Center's investigators found 2,4,5-T products readily available in Washington and Baltimore garden and hardware stores despite DA's claiming to have sent out letters to stop sales. The petition accused DA of being remiss in not suspending 2,4,5-T relatives such as "Silvex," a common weed killer.

On top of all this, now even 2,4-D is suspect, while the surgeon general insists that ill-effects produced in animals are inconclusive in terms of judging the hazard to human health.

In sum, the government officials have not yet interpreted pesticide laws to place the burden of proof on pesticide makers. Test animals may suffer and so may bees, birds, fish, and other animals, but until man falls ill, the official attitude continues that the chemicals can be used with or without strict limitations.

In a memorandum to his department, December 23, 1964, then Secretary of Agriculture Orville Freeman wrote that "When residual pesticides must be used to control or eliminate pests, they shall be used in minimal effective amounts, applied precisely to the infested area, and at minimal effective frequency. Biological, ecological, or cultural methods or nonpersistent and low toxicity pesticides will be used whenever such means are feasible and will safely and effectively control or eliminate target pests."

On a steamy June afternoon in 1969, the present Secretary of Agriculture, Clifford Hardin, handed me a copy of that memo to indicate that it was his policy, too. That may have been his intent, but as of late June, 1970, it had not been backed vigorously or effectively.

WHAT YOU CAN DO

If I appear to have taken up an inordinate amount of space on news developments that have already been publicized, it was for several reasons. Tied together in perspective, these events lead to an inescapable conclusion. Pesticides are potent and potentially harmful beyond our comprehension. No sooner do we think we have an answer than it becomes a troubling question. We focus on DDT and other suspects come to the forefront. We cannot say that just because it is biodegradable or short-lived that a chemical is fit for use. It may be instantaneously lethal or harmful to living things or it may kill pest predators as well as pests or other creatures whose natural role may not have been properly appreciated until the balance was upset. These last few pages have not suggested to you ways of making your crops or garden grow *safely,* but they have linked episodes that might suggest to you that *any* pesticide should be questioned and proved safe beyond a doubt before it is used. Until that rule is established, what basic approaches will help? What are the alternatives?

In General

❧ From the texts cited in the Bibliography, from local conservation and horticultural groups, agricultural schools, even from regional representatives of the Department of Agriculture, find out how you can diversify your planting and arrange the bloom and harvest schedule so as to obtain a balancing mix of pests and their natural predators. The following two examples were reported in *Environment* magazine's special report on pesticides. Farmers in California found that mowing wide, contiguous fields of alfalfa destroyed the balance of natural enemies that kept the spotted aphid in check. Yet when they cut the hay in strips, the predators of the alfalfa aphids maintained a refuge, while the crop matured at staggered intervals. California wine growers asked the University of California for help in overcoming the grape leafhopper which had become immune to organophosphates. It was discovered that the leafhopper had a natural enemy, a tiny wasp (*anagrus*), but this predator occurred too late in the season to prevent damage to tender young grape leaves. Why? Because the wasps wintered in blackberry vines where they had another prey. Since blackberries were chemically treated as weeds in the vineyards, the wasps had to come from patches some distance away, arriving too late to be effective. Now blackberries are being planted experimentally in the vineyards so the wasps can enjoy their summer and winter feasts without long travel between. Diversity invariably pays off in a garden, too. Different plants and trees attract different birds and bugs, setting up a precarious balance of power among your plants and bowers.

❧ Avoid persistent chlorinated hydrocarbon products containing aldrin, chlordane, DDD, DDT, dieldrin, endrin, heptachlor, Kelthane, Lindane, Methoxychlor, ovex, tedion, thiodan, and toxaphene—to name most common formulas. Some are less persistent than others. For example, methoxychlor is often suggested as a substitute for DDT, particularly in combating the Dutch elm beetle that spreads the killing elm fungus, but it is twice as expensive and has not been particularly effective. Nothing has really stopped the Dutch elm disease short of interspersing rows of the stately trees with locusts and other trees to break up the chain-like movement of the beetles. The only answer to this tragic blight that has terrorized town greens in the east and midwest is biological controls, to which scientists presently are devoting a lot of energy.

❦ Choose a selective pesticide, not one that will kill a wide spectrum of insects, and avoid fertilizer-pesticide mixes (double or triple action formulas) that are touted as being able to kill your weeds and bugs while making the grass or plants grow. It is more costly to a pesticide manufacturer to market a selective chemical that has limited use and that is why the stores are full of versatile poisons that can be used widely on different pests in different climates.

❦ Once you have purchased your poison, there are three rules: 1. Do not use it until it is absolutely necessary, until the pest is present. Spraying in anticipation not only may kill predators but start you on the way to exceeding safe dosages. 2. Do not overspray but try to pinpoint your target by learning where the pests hide or do damage and then use as little as you can to do the job. 3. If one application is not enough, seek advice from the sources previously cited as to the minimal number of "strikes" that will bring results.

On the Farm

Since DDT and other synthetic pesticides were marketed, dramatic progress has been made in developing selective pesticides and, more importantly, biological and other natural methods of control. Rachel Carson suggested that parasitic wasps might be imported to feed on Japanese beetle grubs. This approach is now common and there are others such as sterilization, introduction of hormones to cause deformities, and application of pathogenic substances to produce crippling viruses among pest populations. One of the pioneers in this movement is Dr. Edward F. Knipling, Director of the Entomology Research Division of the Agricultural Research Service. In a paper (see Bibliography) he described some new techniques and proven methods of combat. "Nonpoisonous pest control methods are not new," he notes. For example, when the cottony-cushion scale, an accidental from Australia, threatened to destroy California's citrus in the 1880's, the scale's natural enemy was retrieved from "down under" and, fortunately, balance was returned without any new side-effects.

Parasites are being sought and transplanted to control the gypsy moth. New strains of alfalfa and wheat have been developed that resist traditional pests. And a promising new method is the use of "attractants." As explained by Knipling, "Insects respond to various chemical substances in plants in their search for food and to chemicals produced by the insect for mating. They also respond to light, sound, and possibly to radiations not yet identified. Scientists are trying to pinpoint these attractants and to find ways of using them to detect and control insect pests."

In 1937, Knipling came up with the idea of sterilizing screwworms—debilitating infesters of cattle—and releasing them so that eventually reproduction would be reduced to zero because of a chain of infertility being established. It might take five or more generations but the life span of insects is counted in weeks. Knipling did not get a chance to do more than test his theory until 1954, in Curaçao. It worked. It was next tried in Florida, again with sensational results, and now the tactic is used along the Mexican border, although only recently have Mexican authorities agreed to set loose sterilized worms on their side, thus protecting the U.S. program from infiltration.

This is the problem that has prevented application of this method to control boll weevils, cotton pests that are the object of 35 percent of our insecticides and that cost planters $300 million in annual damage (although cornear worms are worse pests, costing $500 million). Boll weevils reproduce every three weeks, so they successfully bounce back, and have even multiplied, from pesticide spraying. But Knipling thinks that if sterilization could be applied massively, or conducted in isolated areas where invading weevils from other fields won't upset the equation, it might take less than ten generations to get rid of the pest. In the spring of 1970, DA did move against another cotton pest, the boll worm, by dropping 800,000 sterilized worms in the San Joaquin and Cachella valleys in California.

※ Certainly, while these natural or biological methods have not been widely applied yet, it is worth inquiring of regional ARS representatives to find out if there are possibilities for such a solution in your fields. The chances are that if they are not available as an exclusive solution, they might be applied as part of what is termed "an integrated program," coupling use of selective chemicals and biological methods. For example, in order to segregate boll weevils, attractants may have to be used to round up local populations of weevils. Chemicals might be applied sparingly to close off the area and then sterilized boll weevils set loose.

※ Beware of salesmen. It is conceded in government circles and is often all too obvious in the field that not only is pesticide performance data incomplete and misleading, but the chemicals are purveyed by manufacturers' representatives who, unfortunately, seldom if ever give a fair explanation of their product's ecological effects. In addition, DA field men, state agricultural agents, and growers' association officials too frequently do not question the material pressed into their hands by salesmen. The man from a major company which markets chemicals arrives and says that the scientists at Shell, Standard Oil, Olin, or Dow have just discovered a new pesticide that will work not only on this bug and this crop but on that nuisance in the other field, so you won't have to spray as heavily or frequently, and so forth. Robert van den Bosch, Berkeley entomologist, wrote in a workbook put out by the Scientists' Institute for Public Information (see Bibliography) that "The salesman is the key to the system, for he serves as diagnostician, therapist, and pill dispenser. And what is particularly disturbing is that he need not demonstrate technical competence to perform in this multiple capacity." This is quite true and is an important issue in reforms being considered in the federal government apparatus. It should be taken up by local governments, too. There are no laws requiring pesticide salesmen to be professionally competent or licensed like pharmacists. "Yet this person," wrote van den Bosch, "deals with extremely complex ecological problems and utilizes some of the most deadly and ecologically disruptive chemicals devised by science." Until the system is made honest, your only recourse is to study texts, question DA representatives unrelentingly, and ask conservationists or university biologists (e.g., agricultural extension service experts) for the other side of the salesman's picture.

The Home Ecologist

"The home environment is surely the place where the utmost caution is in order," wrote Shirley A. Briggs, executive director of the Rachel Carson Trust for the Living Environment, Inc., in a twenty-one page guide provided exclusively for homeowners.

The implications of that statement are grave. They really

go beyond the confines of your property. This was sensitively phrased by Stanford biologists Donald Kennedy and John Hessel in *Cry California*'s special issue on pesticides. "When a home gardener selects a particular pesticide—or when he chooses *not* to use one—he makes an ecological decision," they wrote. "That decision does not apply merely to his own land; it is not in the nature of ecological systems to be subdivided by backyards." So what rules should you follow?

The suggestions for farmers certainly apply to the home, although many hazardous chemicals are not sold for household use but are to be used only in supervised programs. A more important distinction is that *your garden criteria should not have to include economic profit.* Food standards may make farmers go overboard to keep the last insect or cornborer out of their produce but you should not feel compelled to worry about these esthetic detriments, which rarely are hazardous to health.

※ Landscape and garden with as much diversity as possible, and try to plant only native flora—not outside varieties that may seem well suited to the climate—unless a visiting species has been proven successful in helping to achieve a pest balance.

※ If you have thus created or maintained a garden setting in which the diversity of flora influences a tolerable balance of pests and pest predators, you shouldn't need pesticides. In addition, if you make your own compost by piling dead leaves and clippings into a bin, then you may get by without any chemicals on your plot. However, if it seems necessary to purchase a pesticide, don't buy one that will kill the beneficial insects, particularly the bees that perform an invaluable service by pollinating. And, as Shirley Briggs warns, "Beware of any product that promises to solve almost all your problems with no harmful effects. We still do not get anything for nothing."

※ As already emphasized, do not spray before the pest is a problem, use the least amount to do the job, and minimize applications. Often a hard, fine spray of water alone is enough.

Many gardeners swear by it and I will attest that it works on aphids. It also keeps down leafhoppers and spittlebugs. It is also worth finding out from local horticulture and conservation groups or university experts whether biological controls are available for your home pests. Milky spore disease, for example, can be instigated to cope with Japanese beetle grubs in your orchard. Avoid highly toxic inorganic chemicals containing arsenic and fluorine but you do not have to shun so-called "botanical" insecticides made from plant substances that mother nature gave certain plants to defend themselves. Examples are nicotine sulphate, pyrethrum, and Rotenone. Again, avoid the two-for-one packages and try to pick a chemical that is selective, not lethal to a wide variety of nuisances.

※ Do not apply chemicals when the wind is blowing or about to blow—they might become your neighbor's problem. It also pays to learn the natural characteristics of your property, particularly how water drains and *where it runs off to.* If you have a well system, you certainly do not want to contaminate your drinking supply. Neither do you want to pollute a stream that runs through your place. Remember, one chemical may not damage aquatic organisms directly, but acting synergistically with other chemicals and fertilizers it could endanger the waterway. In sum, know your ecosystem, the interplay between plants and wildlife.

※ Organic farming and gardening has increasingly become an intriguing challenge. Relying on natural factors, turning your own compost bins, having a duck to go after slugs and attracting the kinds of birds and bugs that devour pests, really tests your ingenuity. At the risk of sounding like a fuzzy-headed moralist, I contend that such a challenge will also give you a far greater feeling of identification with your natural surroundings. You can boost the natural process and add a few old remedies. For example, stock your compost pile or topsoil with earthworms available at most fish bait outlets. They will do wonders for the earth, aid organic decomposition by mixing dead plant matter with the soil, and loosen the ground so that it absorbs water readily and evenly. Then you can try diesel oil *very precisely* on weeds. It will clog their pores, thus preventing them from transpiring.

⚜ The following is a guide to garden chemicals prepared by Shirley Briggs for the Rachel Carson Trust and is reproduced here with permission. She insists that since new data constantly turn up, she is always revising this advice. Some of her stipulations with regard to the following table (done in the winter of 1970) I have included in an addendum. If you want to be certain of what to avoid and what to use and would like to know more about pesticide disposal, I suggest that you write to the Rachel Carson Trust, 8940 Jones Mill Road., Washington, D.C., 20015, and ask for their new pesticides pamphlet that was being worked on as this book was written. It will include a list of trade names corresponding to common names, a most helpful contribution, since the pace of chemical production results in a constant, baffling flow of new brands.

GUIDE TO GARDEN CHEMICALS (RACHEL CARSON TRUST)

Some Acceptable Pesticides

While these include some potent poisons, they do not persist in food chains or in the environment generally, and they have not been shown dangerous to wildlife except as noted. They are fairly selective in their targets, and will not spread indiscriminately through the environment. Most have been used for a long time, and their effects are generally well known.

Some of these, as with any pesticide, can be irritating to people with allergies. Avoid any large-scale spraying and never spray or dust on windy days. Keep applications as local as possible.

Bordeaux mixture (copper sulphate-calcium hydroxide) if not fortified with lead or calcium arsenate	Fungicide and insecticide
Copper-lime mixtures	
Pyrethrum (Allethrin is a synthetic analogue)	Be careful of applications that could lead to human allergic reactions.
Rotenone (derris, cube)	Toxic to fish; avoid applications that could drain into ponds or streams.
Ryania	
Sabadilla	Toxic to bees, but not in the usual low dosage
Sulphur, lime-sulphur	
Nicotine sulphate (Use with caution.)	Highly toxic, but non-persistent. Has not been shown dangerous to wildlife, but must be handled with great care to avoid human poisoning.

BACTERIAL INSECTICIDES

Bacillus thuringiensis Berliner	Specific for a variety of species of moths and certain other Lepidoptera
Milky spore disease	Specific for Japanese beetle grubs

DESICCANTS

Silica aerogel in brand names Dri-Die or, with pyrethrum, as Drione	A desiccant powder, especially for use in the house, as it is nontoxic to humans or pets, though it can be drying if inhaled much. Effective on insects because it penetrates their waxy exoskeleton. This particular kind of silica aerogel absorbs lipids and is thus most effective.
Diatomaceous earth	Also kills insects by contact and desiccation

RODENTICIDES

Anticoagulants, such as Warfarin	Selective for rats and mice only if they have a chance to get repeated doses. Use a locked bait-box which children and pets cannot get into.
Red Squill	

Some Pesticides to Avoid

PENSISTENT, BROAD-SPECTRUM PESTICIDES

These remain in the ecosystem with little change for many years, accumulating in soil and organisms, and tending to concentrate in animals at the end of food chains. They are toxic to a wide range of organisms, and cannot be confined to the target species once they are released in an environment.

CHLORINATED HYDROCARBONS

Aldrin
BHC (benzene hexachloride, Lindane)
Chlordane
DDD (TDE)
DDT
Dieldrin
Endrin
Heptachlor
Toxaphene

Act as nerve poisons, and are stored in the fat of animals; can be released suddenly into other tissues when stress or reduction of food intake depletes fat reserves, thus reaching critical areas like the brain long after initial exposure. They are synthetic compounds, not ever found in nature before their wide use since 1946. Have little-understood metabolic and other physiological effects. Can disrupt soil ecology and are very persistent. Some give evidence of carcinogenic effects.

Methoxychlor is less toxic than the others to warm-blooded animals but it has known hormonal effects, with estrogenic capabilities. Carcinogenic effects are considered very low in normal use. Toxic to fish.

GENERAL PROTOPLASMIC POISONS

Arsenic, in many forms (used as insecticide, herbicide, rodenticide)

A cumulative poison in animals and soil; can make soil sterile to plants and soil organisms. Carcinogenic.

Mercury, organic and inorganic

Increasingly used as a fungicide, often on golf courses and cemeteries. Has caused a variety of problems with wildlife and food supplies in Sweden, where it has been much used in pesticides.

Dinitro compounds

Dinitrophenol (DNBPH, DNOSBP)	Cumulative poisons, used as fungicides, insecticides, miticides, or herbicides. Very toxic, can be absorbed through skin.
Pentachlorophenol	Action similar to the dinitrophenols. Also used as a bactericide. Symptoms of poisoning similar to those of organic phosphate poisoning, but treatment must be different.

RODENTICIDES

Alpha naphthythiourea (ANTU)	Evidence of carcinogenic effects.
Phosphorus, white or yellow Sodium fluoracetate (1080) Thallium	Extremely toxic; poisons with these active ingredients are not registered for home use except for phosphorus, which may not be in or on anything resembling human food.

NONPERSISTENT, BROAD-SPECTRUM PESTICIDES

These are new synthetic chemicals, as are chlorinated hydrocarbons. They include highly poisonous compounds, but may have less total effect on the environment than the preceding kinds because they break down into nontoxic materials fairly quickly. Their indiscriminate effects make them undesirable for home or garden use.

ORGANIC PHOSPHATES

Chlorthion
Demetron
Diazinon (trade name)
EPN
Parathion
Phosdrin
Schradan
TEPP

A group of chemicals originally developed as chemical warfare agents. Toxic chiefly through cholinesterase inhibition, and resultant breakdown in nerve and muscle response. Some may show potentiation: combinations with other chemicals may be many times more toxic than the original components would indicate. Sublethal effects and full impact on the environment not well known, as is also true of the chlorinated hydrocarbons.

Malathion is usually listed as much less toxic to warmblooded animals than are the other organic phosphates. In home use, acute poisoning is possible from malathion, however, and severe kills of birds and other wildlife have resulted from indiscriminate spraying with it.

CARBAMATES

Carbaryl

Methyl carbamate herbicides

Carbamates are still under study, and while evidence so far does not indicate serious danger in their use, some doubts have not yet been dispelled. They are not cumulative, act through cholinesterase inhibition, and are now considered of medium toxicity to humans. They may be very toxic to bees.

FUNGICIDES

Captan

Folpet

Recent experiments with animals show high incidence of birth deformities following exposure to these, whose chemical structure is similar to that of thalidomide.

Herbicides

Growing suspicions about health hazards from the herbicides formerly thought safest have led to the recent restrictions on 2,4,5-T, and there are doubts about the similar 2,4-D. Any chemicals so physiologically active with one form of protoplasm should have had more study than these received before being released widely into our environment. Other herbicides on the market are suspect as carcinogens, mutagens, or may interfere with embryological development. Older types, such as arsenic, have the added hazards of persistence. In the home environment, where no pressing reasons exist for chemical controls of this sort except as a convenience, cultural methods to destroy unwanted plants are thus much to be preferred. Competent botanists have used 2,4-D, 2,4,5-T, and Delapon sodium salts effectively, and relatively safely, in the very selective control of vegetation, applying the chemicals precisely to the vulnerable parts of individual plants. Overall spraying or distributing of herbicide-containing materials is not recommended by these experts. It is often not much harder to remove such plants by other means.

As usually applied, herbicides can create serious problems with drift, killing plants not intended. Even on calm days, sprays of these materials can be detected surprising distances away, and damage may occur a mile or more from point of application. Danger can also arise from the use of lawn conditioners which contain herbicides, such as 2,4-D and 2,4,5-T in granular form. In hot weather these can volatilize and affect susceptible plants nearby. Your neighbor's lawn treatment may deform your roses, grapes, or tomatoes, especially if he has overdone the dosage.

Federal regulations for farmers have prohibited concentration of 2,4-D at more than two pounds per acre, and the state of Iowa limits this to one pound because of the volatility problem. Homeowners should practice converting the directions for their garden chemicals from the usual "per 1,000 square feet" formula to the "pounds per acre" rule applied to farm use. There are no federal regulations governing amounts permissible in home and garden use, so alarmingly high rates are sometimes recommended. Some weed killers marketed for home use have contained arsenic compounds at a proportion sixty times that permitted on the farm.

Even more problems can arise from the use of "weed bars," which are usually 2,4-D and wax. Placing or dragging these bars near such plants as roses, grapes, or petunias can be damaging. Herbicides have been considered less dangerous than the more persistent chemicals, since they break down fairly soon in the environment, but once plants have been exposed to them no remedies are known, and you can only hope the harm will not be fatal.

Disposal of Pesticides

Empty containers and surplus pesticides must be disposed of very carefully. The persistent pesticides, notably DDT and the other chlorinated hydrocarbons and some other compounds take many years to break down into less toxic material. Arsenic, lead, and selenium remain toxic indefinitely. For these especially, disposal should assure that they will not later escape into the environment. No really sure way has yet been found to do this. Burning in specially designed incinerators that will maintain "red heat" temperature of 700°C (about 1290°F) and preferably one that has an after-burner, will destroy all organic material. Very few municipal incinerators meet this test. Burial requires knowledge of the terrain and of future land use that few people can guarantee. Until research into new disposal methods provides us with better answers, each of us must weigh possible methods and choose the best under the circumstances.

Some Rules and Possibilities

❧ Never put pesticides down the drain, into the trash can, or on a public dump unless it has special provision for pesticide disposal. This means deep and immediate burial where the materials will not reach any groundwater or surface water and where disturbance of the soil will not occur to unearth the pesticides later.

❧ Old containers must always be disposed of as carefully as the poisons themselves. Never reuse them and do not leave them in usable shape for others to use, even though you bury them.

❧ Burial on your own property should only be done when

the materials can be put into a hole well away from trees, desirable shrubs, or plants, and at least fifty feet from any well or surface water and not too near a home. Be sure they will not leak down into the groundwater. The hole must be deep enough to cover the pesticides with at least three feet of soil immediately. (The question here is whether people will ever dig so deep a hole.) Put aerosol cans in intact. If the highest possible water table is less than five feet below the surface, do not bury pesticides there at all. Chemicals in glass or metal containers will presumably retain their toxicity indefinitely if buried intact, so some authorities recommend breaking such containers so the contents will spread out into the soil and be subject to natural degrading over the years. The decomposition of certain organic pesticides can be hastened by mixing them or surrounding them with at least ten times their weights of quicklime or hydrated lime. Your estimate of the soil condition might indicate whether this will be desirable in your case.

❧ Some states have set up collection points from which surplus pesticides will be taken for careful disposal. If your state or county has no such service, suggest that it be arranged, and check on the thoroughness with which they take proper precautions.

❧ Organic phosphates, which decompose rapidly, are more suitable for burial, but the same precautions should be taken. Admixture with lime is particularly effective with these materials. Later land use need not be such a problem, though.

❧ Public health officials should be aware of the problem, and may be able to suggest solutions for your area. Be sure they are really familiar with the problem.

❧ Herbicides should never be burned. Most organic herbicides degrade fairly rapidly, but those containing arsenic compounds will remain toxic indefinitely. Bury all herbicides far from any vegetation that might be affected. Burning releases very toxic fumes, and some herbicides will explode.

꙰ Local incinerators seldom burn at high enough temperatures to destroy pesticides completely, and they may release poisonous fumes. If the local authorities are willing to make special provision for burning your quantity of pesticides or containers, wrap them in several layers of newspaper to assure as high a heat as possible. The package should be placed beneath a layer of combustible materials so that escaping vapors will be forced to pass through a zone of flame.

꙰ Large quantities of pesticides present a greater problem. They must be buried at least six feet above the highest groundwater level, where the land is not apt to be used for crops, pasture, or building sites, and at least three hundred feet from any wells or surface water. They must be covered immediately with at least three feet of soil.

꙰ The average city or suburban property should not be used as a disposal site under most conditions, partly because the pesticides can rarely be buried very far from a house, and also because future land use is unpredictable. It is also unwise to centralize disposal too much, as at a dump, because serious contamination could result if the land is disturbed years later.

Addendum

Shirley Briggs does her best to keep a poison-free garden around her own home. Sometimes she uses a botanical. For bothersome, repulsive slugs she uses plain old beer set out on the lawn in a shallow bowl or saucer. It attracts the slugs, which then drown in the beverage. Since the preceding guide was drafted, she has developed hard misgivings about Sevin, the trade name for carbaryl. It is tertogenic in animals, meaning that it produces congenital defects like Thalidomide. And yet, as will be pointed out in a vignette at the end of this chapter, Sevin is generally recommended *despite new warnings by HEW*.

"Arsenic in any form whatsoever should not be used," she says, adding that "There is just no way of getting rid of that stuff. It survives even the high temperatures of incinerators." She also feels more strongly against Captan and Folpet, fungicide powders used on fruit and vegetables. The Nixon administration report cited studies showing that these chemicals had produced birth deformities in mice and chickens.

LEGAL REFORMS

Everyone who does not have a vested interest in pesticides seems willing to accept the need for legal or institutional reforms to control these chemicals, just as dangerous drugs are tightly regulated. There are proposals pending in Congress that would tighten registration and inspection procedures and restrict marketing and use of certain kinds of chemicals to trained specialists. Wisconsin's Gaylord Nelson has proposed a pesticides commission to conduct continuous review of pesticides' content and regulation. As this chapter was written, the Nixon administration proposed that a single federal department, the Environmental Protection Agency, administrate air and water quality programs and pesticide and other environmental health activities. Assuming that Congress would approve this plan, it would be some time before present programs were reassessed and then enforced to a much greater degree than before. It was good to see pesticides regulation removed from DA but that did not necessarily mean that tougher action was immediately forthcoming. Government reorganization is laudable if a sensible purpose is served but it also usually results in temporary disorganization and delay.

State laws that follow federal guidelines are usually inadequate, although some states such as Arizona, California, Michigan, Massachusetts, and Wisconsin have passed or are considering tough laws against persistent pesticides. There are no rules forcing manufacturers to publicly disclose more about their potions—if only information concerning possible environmental side-effects. County agricultural agents are thus at the mercy of the chemical hucksters, and the farmers who depend on these advisers are really caught in the crossfire.

Present federal farm programs encourage "monoculture" farming and high yields. Thus, a farmer is given no incentive to diversify his crops to keep a pest-predator balance and he is prodded to produce as much as he can of some crops (while forsaking others) with no admonition that this increase not be at the expense of the environment. In the Midwest states, farmers are told they can plant earlier in the spring and seed a bigger area if they dig drainage ditches around their fields and also dry out their marshes and wetlands. Not only are important predator habitats destroyed but the ditches carry runoff loaded with hazardous chemicals into streams and rivers. I was told emphatically by Minnesota and North Dakota farmers a year ago that they were given no alternative to these practices. Indeed, in the Washington bureaucracy that is judged to be the case.

Pesticide reform should include solving two related problems: the trend to broader gauged pesticides and the setting of higher food standards. Registration laws ought to require far more selective chemicals, even if government research and development programs have to be incorporated to produce them. Finally, our expectations are too high on food. We want corn that has not one insect trace in it, and the government requires that foods have no insect evidence at all, even if it is harmless. Certainly the home gardener is better off with a few cornborers than an ear full of pesticides. As the Stanford biologists Kennedy and Hessel put it, "In actual fact, the chemicals we add to our produce while it is still growing often merely substitute potentially harmful pesticide impurities for harmless insect impurities.

AUTHOR'S BRIEF

In the spring of 1970 the seventeen-year locusts attacked. I was prepared by local newspaper accounts not to worry. Unlike the voracious desert locust that lays waste African and mideast countries, this cicada would do no more than kill tree branch tips by burdening them with its eggs—an example of "nature's pruning." Only young trees might die from this onslaught, their slender shoots so destroyed that they could no longer transpire.

But that was precisely the threat to our new house in the middle of a field in Virginia. We had recently landscaped it with many saplings of dogwood, locust, gum, oak, and cherry and other fruit trees. For a while, the locusts stayed in the woods, making a distracting high-pitched whirring noise like ambulances or, according to my ten-year-old son, like a flying saucer landing. But in their waning moments, before they would die and abandon the tree tips which would finally drop to the ground, planting cicada larvae for the crop of 1987, the adult locusts moved in. We anguished.

A call to two experts produced the same advice. Buy a brand of carbaryl called "Sevin," they said. It's a nonpersistent powder and apply it at dusk where there is no wind and the dew will hold it on the leaves. So done. And even though the locusts seemed to die naturallly over a few days—not from Sevin—I was reassured to find it approved in a guide published by the National Audubon Society.

Then two unsettling things happened. While driving I saw that a farmer had protected his young fruit trees by veiling them with strawberry netting that seemed much wiser than our approach. And then as I researched this chapter, I found that some biologists question Sevin. It is toxic to bees. Is has caused malsymptoms in animals.

Thus there is no way I can be *sure* of safety in my garden as long as I use chemicals. Pity the birds, I thought. So we have returned to habits previously followed out of laziness as much as diligence—no pesticides.

A woman outside San Francisco named Miss Tilly was interviewed by a friend of mine, Richard Reinhardt, who wrote about her poison-free garden for *Cry California*.

"If the birds will eat the bugs," she said, "we'll spare them a little fruit. When this garden was started, we didn't hear about all these chemicals, yet the garden grew." That it did, as Reinhardt reported. It was a lush tangle of growth that amazed horticulturists, even though the pests managed to get in some licks. "I'd rather have a few bugs in the garden than to have a garden full of poisons," was Miss Tilly's answer.

10 ❧ ENVIRON-MENTAL LAW

DOES THE UNITED STATES Constitution guarantee you a right to environmental quality? Federal and state agencies are supposed to protect your air and water and local governments traditionally have regulated the use of land, but what if they fail to carry out their mandate, as is so often the case? Or what if a conflicting interest prevails, as with pesticides? In a society that stresses innovation and "progress," how can technological growth be channeled to prevent adverse effects on the environment? As neatly put by James W. Moorman, a young Washington lawyer who has helped carry the fight against DDT and the Alaska pipeline, "The question is what type of evidence can be introduced against the values of technocracy. How does one present an objective analysis of the degree of scenic beauty of a specific place when asking the decision maker to elevate rather than fatten mankind?"

For years, conservationists have gone to bat in the courts with two strikes against them. The guarantees and protections they seek are, to be sure, most obvious. A citizen wants relief from automobile and bus exhaust, smoke belching from a factory smokestack, the noise of jets overhead, and the bulldozer tearing up a piece of wild land or filling a marsh. But, however obvious and precise these grievances are to the beleaguered citizen, they are vague in the eyes of the law.

In the first place, the citizen cannot just go to court and complain. He has to relate his complaint to a specific local, state, or federal statute. Then he will be challenged because he has not claimed a specific economic deprivation and therefore lacks standing to sue. If he gets over that hurdle he finds that *he* has the burden of proof. That is, he has to show that the factory, or the jetport, or whatever, is damaging him, rather than the more appealing situation in which the polluter is required to show that its actions are *not* doing damage. So, there stands the citizen in court, having related his complaint to a specific statute, having claimed economic deprivation, and bearing the burden of proof. His next problem is that the scientific data he needs to prove his case are probably considered "privileged information" and are locked away in the files of the offending organization or government bureau. The citizen cannot get it, so he has to develop his own scientific data, or go home.

Until recently, if you asked a lawyer who has struggled against these odds to cite the legal precedents that would help him, he would count them on the fingers of one hand. To be sure, there have been countless nuisance cases and private damage suits, but these haven't established broad principles. So the field is wide open, confused, and tough.

With these problems in mind, forty-six conservationists and lawyers, joined by "observers" and several reporters, met at Airlie House in the northern Virginia countryside in the fall of 1969 to tie up the few loose strands that existed in the field of environmental law and to develop new avenues for future citizen action.

NO MONEY AND
NO STANDING

The conference began on a note of despair. David Sive, a New York City lawyer, lamented that "In environmental litigation, we don't have the money and we usually don't start litigating until the bulldozer is overhead." Sive and others pointed out that conservationists—as individuals or in groups— are not concerned with particular properties in which they hold direct stakes. Their interest is broad and diffuse, extending from the unspoiled reaches of the Arctic, now threatened by oil exploration, to the rugged Maine coast, where the refineries hope to locate. They are aroused and held together by crises, unorganized to meet a threat until it is visible and only then able to stir up sufficient indignation to raise funds for the fight. The offending developer or agency has had a long headstart with plenty of ready cash for advertising and teams of legal and scientific experts putting together learned, if not slick, rebuttals to every conceivable attack. And then just getting into court to be heard has until recently been difficult. There is no law allowing a single citizen to bring suit against an environmental destroyer unless he personally and directly is victim and then his case is weak. And, invariably, conservation organizations have had to spend an inordinate amount of time and effort in establishing the fact that *they* have a right and a claim when, for example, a piece of scenic land is being stripped of natural assets that make it generally beneficial to the public.

A group's standing to sue over a conservation issue was first established in the case of the Scenic Hudson Preservation Conference v. the FPC in 1965. Sive participated in that suit in which citizens sought to protect Storm King Mountain on the Hudson River Gorge from being developed as a pumped storage hydroelectric facility. The plan of the giant utility, Consolidated Edison, was to use conventional generators during the nighttime slack period to pump water to an artificially built pond atop the mountain and to release the water back to the river, generating electricity enroute, during the peak daytime demands for power. The conservationists sought to force the Federal Power Commission to give full consideration to environmental factors, mainly aesthetics, when granting Con Ed a construction permit.

The court rejected the FPC's argument that the Scenic Hudson group could not obtain judicial review because it claimed no economic injury. In addition, the court ruled that the FPC Act should be construed to protect such noneconomic issues as "the aesthetic, conservational, and recreational aspects of power development." Aesthetic quality, of course, remains difficult to define. Unfortunately, Supreme Court Justice Douglas was writing a dissenting opinion in another case (Berman v. Parker) when he contended that "It is within the power of the legislature to determine that the community should be beautiful as well as healthy, spacious as well as clean, well balanced as well as carefully patrolled. . . ." But a decision since Storm King (Udall v. FPC) has expanded the definition of "recreational aspects" to include the protection of wildlife such as anadromous fish.

A new hearing was held as a result of the Scenic Hudson challenge and Con Ed was able to satisfy FPC criteria by planning its installation underground. But the battle continued as this chapter was written, the ranks of the opposition swelled by addition of the City of New York and the State of Connecticut, and the suit pressed not only for aesthetic reasons but on the grounds that the river would be used so as to endanger fish populations, and the underground facility would threaten the city's Catskill Aqueduct.

THE LESSONS OF SANTA BARBARA

It took a disaster that was all too visible to galvanize the so-called "new conservation" movement and spotlight, unlike any previous U.S. pollution crisis, the unsolved problems of environmental liability. The event influenced the passage of new legislation (see Chapter 8). It was the Santa Barbara oil spill in mid-winter 1969, and in the opinion of Malcolm

Baldwin, a legal associate of the Conservation Foundation and an organizer of the Airlie House meeting, it "stands as a classic illustration of how the environment is mismanaged by public agency and private business decisions." His paper on the subject was the opening topic of discussion at Airlie House.

The federal government was under tremendous pressure to lease the submerged lands of the Santa Barbara channel, which were estimated to contain around four billion barrels of oil, not much less than our annual consumption of 4.7 bbls. The President of the United States, Lyndon Johnson, and his Budget Bureau were itching to have the revenue that the channel leases would gross ($603 million).

There was no compelling reason to question the safety of oil operations in these waters in view of the record of companies in the Gulf of Mexico (since tarnished by Chevron Oil Co.) and the fact that their scientists insisted there were no ecological dangers or seismic hazards. Moreover, conservationists mainly were concerned with the aesthetic intrusion of drilling platforms and had already persuaded both the State of California and the federal government to create a marine sanctuary in front of the town and harbor of Santa Barbara so that the view of the Pacific horizon and the Channel Islands was clear.

But negotiations were cozy and conducted behind a veil of secrecy. Federal drilling regulations did not require the oil companies to reveal the data they used to support their claims of safety. The material has always been treated by the companies as proprietary information that would put them at a competitive disadvantage if revealed. Federal officials were sufficiently *lulled* by Union Oil Co.'s assurances to give that firm drastic variances on drill casings required in sinking the well that later erupted on platform A. *After* the damage had been done, Union officials conceded that it probably never would have occurred had the routine safety precautions been taken on the oil rig. And William Pecora, Director of the U.S. Geological Survey, the agency that supervises offshore drilling, said what he must have known all along, that "without any question of doubt, we did not have as much information on the detailed reservoir conditions as existed in the oil company files."

Drilling in the immediate area of platform A was ordered stopped while a Presidential panel investigated the accident as well as potential hazards throughout that channel, but it was eventually allowed to continue under stiffer casing requirements and other regulations. Another oil company filed a claim against the U.S. after the 1970 Water Quality Act imposed absolute liability for cleaning up offshore spills. Including county and state suits, damage claims against the U.S. exceeded one billion dollars. Thus it is likely that had the spill been anticipated, leasing in these waters never would have been permitted.

Even after this lesson, however, it is questionable whether such tragedies can be prevented. Victor Yannacone, the colorful lawyer who represented the Environmental Defense Fund in its early battles, recounted at Airlie House how he had investigated the Santa Barbara spill and advised against pressing a suit to stop all drilling, because of both the high cost of pressing such a suit and significant scientific disadvantages. He said that to obtain at least a temporary restraining order, he was "interested in only one question," whether the act to be complained about would "cause serious, permanent, and irreparable damage." Aesthetic grounds were pointless, and one month after the oil had begun to roil to the ocean surface he was still unable to elicit the "proprietary" scientific data from the oil companies. The only seismologists who really knew what geological hazards existed in the ocean floor were in the pay of the companies. The only tactic remaining, said Yannacone, was to demand that the oil companies show cause as to why they should continue drilling. This might flush out scientific secrets, but then what if this information showed that drilling was safe? A lot of time and money would have been wasted.

The new drilling regulations, issued after Yannacone had decided not to become involved in Santa Barbara, required the Department of Interior to take environmental factors into account when issuing well permits. At that point Yannacone might have recommended bringing action under the new rule (Reg. 3381.4), forcing DI's hand. However, as later expressed in another paper by Baldwin, "It is doubtful that any agency directed to obtain economic benefit from a federal resource will take a strong environmental position where significant immediate revenue reductions would result."

Most of the questions raised by the Santa Barbara spill remain unanswered. It is possible that the new regulations and rules of absolute liability will make offshore operators more careful in the future, and conservationists may gain strength from two new tools that have been applied in recent successes. The first has existed for a long time and, in Baldwin's view, might have been tried before the Santa Barbara crisis. It is the 1899 River and Harbor Act (see Chapter 8) that grants the Army Corps of Engineers authority to deny *construction permits* in navigable waters and was broadened in 1967 to take into account effects on fish, wildlife, and recreation. The second is the 1969 National Environmental Policy Act that requires environmental studies and a statement by federal agencies concerning any federally assisted or licensed activity that might have adverse ecological effects. These two powerful tools will be discussed later in the chapter.

CONSTITUTIONAL RIGHTS

Santa Barbara showed how conservationists are handicapped in shifting the burden of proof and in obtaining the disclosure of information they need before they can even weigh their chances of proving that an environmental threat exists. The event also revealed how the so-called "administrative decisions" of government agencies are essentially arbitrary, as long as they are based on *some* rational alternative, and unresponsive, since after both the Presidential panel and the Department of the Interior studied drilling in the channel they allowed drilling operations to resume.

Looking for protection under the U.S. Constitution has so far not proved productive either. While Yannacone has referred to the equal protection clauses of the Fifth and Fourteenth Amendments and then the Ninth Amendment in entirety, no decisions in EDF's favor have cited these constitutional guarantees. The Ninth Amendment would appear to have the most promise. It reads in full thus: "The enumeration in the Constitution of certain rights shall not be construed to deny or disparage others retained by the people." Both Yannacone and Cornell Professor E. F. Roberts maintained at Airlie House that the public retains the right to a decent environment. "We merely need a ringing decision to ratify this existential fact of life," Roberts wrote in a paper presented at the meeting. Nuisance laws, he argued, do little more than sanction the bargaining away of ones environment. A citizen is simply paid off in damages for noxious air drifting into his backyard or noise assailing his ears and then he is expected to live with the nuisance. But so far, no court has ventured to make the constitutional declaration sought by these lawyers. The consensus among those at Airlie House seemed instead to be summed up by James Moorman, who urged that "Rather than attempt to push Constitutional rights that the courts are not yet ready to accept, we should concentrate on standards that Congress has written into various acts and try to put some life into them."

PUBLIC TRUST

There is one age-old principle that cannot be ruled out of the immediate future, particularly since it has been applied by states (e.g., Georgia and Oregon; see Chapter 8) to protect the coastal zone. It is the doctrine of public trust, and it was advocated at Airlie House by Washington lawyer Anthony Roisman, whose partnership of Berlin, Roisman, and Kessler is in the rare business of taking *only* consumer and environmental protection cases that large established firms handle in small doses, *pro bono publico* (for public service).

The public trust doctrine traditionally has been used to uphold the public's right to enjoy waterfronts and the shallow water over shelves of submerged land. "The trust is the assurance to the people," wrote Roisman in a paper, "that, at any given time, the uses to which property will be put must be consistent with the public interest." He feels that today not only does trust govern the use of public lands (one third of the nation) but that air and water quality legislation has created a presumption that *all* land is subject to restrictions to prevent its abuse at the public's expense. "When Lewis and Clark took their expedition there was an entirely different set of considerations that would have gone into determining what was proper use of the land in this country," Roisman said in discussion. "Today I think the situation is fairly clear because we are virtually running out of land. . . . There is psychological damage created by the inability to find a peaceful place, one without a highway or power land to disrupt it."

A decision cited by Roisman to support his theory has also been hailed as a benchmark in shifting the burden of proof to a developer. In this case, Wildlife Preserves v. Texas Eastern

Natural Gas Co., the New Jersey Supreme Court ruled that the utility would have to show that by putting a gas pipeline through a private marsh refuge—Troy Meadows Preserve—it would not create serious ecological damage and that feasible alternative routes were not available. Before going to court the utility had done neither. Sent back to a lower court, the gas company eventually presented new evidence to satisfy the environmental criteria, at least in the court's judgment. However, the burden of proof and public trust principle was established. Many lawyers, like Michigan Professor Joseph Sax, feel that such precedents need the force of legislation and, accordingly, he helped draft a recently passed law in Michigan that would not only pin the burden of proof on developers but would protect "the public trust in the natural resources of the state" (the Michigan law also gives a private citizen standing to sue in the public interest.)

ACTION-FORCING

Such a law in effect was enacted in Congress to guide the plans and projects undertaken or aided by federal agencies. As already noted, the 1969 National Environmental Protection Act (NEPA) ought to be the lever that James Moorman and others have been counting on to make the federal government environmentally responsible *and* responsive.

National Environmental Protection Act

Some of the key wording in NEPA is cited in a petition filed with the AEC in the summer of 1970 by Roisman's firm on behalf of the National Wildlife Federation, The Sierra Club, and a citizens' group, all of whom sought to halt the construction of a nuclear power plant at Calvert Cliffs on Chesapeake Bay. While the Baltimore Gas and Electric Co. said it followed guidelines in a general report on the effect of thermal discharges on the bay's marine ecology (see Chapter 8), no study of the potential effects of Calvert Cliffs alone was made *before* even construction of the plant was permitted. The petitioners asked the AEC, which licenses nuclear plants, to order BG&E to prepare an environmental statement and, in addition, to begin its own studies of the ecological impact of the power plant and the alternative possibilities.

The petition accurately paraphrased section 102 (c) of NEPA that requires involved federal agencies to consider the following:

"(i) the environmental impact of the proposed action,

(ii) any adverse environmental effects which cannot be avoided should the proposal be implemented,

(iii) alternatives to the proposed action,

(iv) the relationship between local short-term uses of man's environment and the maintenance and enhancement of long-term productivity, and

(v) any irreversible and irretrievable commitments of resources which would be involved in the proposed action should it be implemented."

On May 12, 1970, the President's environmental quality council published in the *Federal Register* the guidelines it planned to follow under NEPA, the bill that created this council. Significantly, Section 11 of the guidelines said that "to the fullest extent possible," NEPA's 102 clause would be applied to projects that had been started *before* the act went into effect. "It is also important in further action that account be taken of environmental consequences not fully evaluated at the outset of the project or program," said the guidelines. "In particular," said the guidelines (Section 2), "alternative actions that will minimize adverse impact should be explored and both the long- and short-range implications to man, his physical and social surroundings, and to nature, should be evaluated in order to avoid to the fullest extent practicable undesirable consequences for the environment."

NEPA, notably section 102, looks like the environmental lawyer's master key. The possibilities for citing it are limitless, since federal agencies are involved in mineral rights and timber leasing, dredging and filling projects, highway construction, power and gas line routes, bridge construction, river dams, and a host of public works activities that have historically been associated with environmental rapine.

The bill has already figured in a monumental court order that delayed the construction of a 390-mile road to haul supplies from Fairbanks, Alaska, to the booming oil operations in Prudhoe Bay on the Arctic Ocean. Of additional and perhaps greater significance, the court injunction prevented the oil companies' consortium, the Tran Alaska Pipeline System (TAPS) from beginning construction of a pipeline in a right of way alongside the roadbed. Moorman, representing one of the conservation groups bringing action, contended that, in view of grave warnings about the pipeline's hazards by the Geological Survey, Interior Secretary Hickel had not complied with NEPA. Furthermore, Moorman argued, the Department of the Interior was in violation of the Mineral Leasing Act of 1920 by allowing a right-of-way for the pipeline almost twice as wide as the fifty-four feet permitted under the act. Since the road and the pipeline were tied together in one project, the lawyer contended successfully that the road could not be built until the pipeline route and design had met (or failed) environmental specifications.

DDT Decisions

The main weakness of NEPA was shored up by two decisions on petitions filed to ban DDT. The law only makes federal agencies *consider* alternatives. So in all likelihood, an agency would follow the Santa Barbara example and do as it pleases. However, in late May, 1970, considering two petitions to stop the use of DDT, the U.S. Court of Appeals in the District of Columbia ruled in effect that (1) the burden of proof remained with the federal agency whose decision affected the health and safety of the environment; (2) that administrative decisions *are* subject to judicial review, and (3) that agencies must provide a record of conversations, correspondence and other scientific deliberations on which they based their decisions. This means that if a federal bureau did not choose the best course of action from an environmental standpoint, conservation litigants could bring action to have this decision reviewed in court and bring into the open *all the data* considered by the federal agency, material previously considered secret and proprietary or not even recorded.

As noted, there were separate DDT petitions. One, filed against the Department of Health, Education and Welfare, asked that zero tolerances be set for DDT in foods, just as is done under the Delaney Amendment (see Chapter 9) for any substance associated with cancer. The court decision, written by Judge J. Skelly Wright, ordered HEW to publish in the *Federal Register* the petition of five national groups which proposed zero tolerances, thus forcing an administrative reassessment of DDT's safety, *with the burden of proof on HEW*. If DDT tolerances were eventually allowed to be continued, Judge Wright wrote, the secretary of HEW would "be required to explain the basis on which he determined such tolerances to be safe." A footnote in the opinion is of special fascination because it gives conservationists an opening to force *reconsideration* of environmental safety *after* a decision has been approved. The judge wrote that "the fact that the present petition seeks revocation of an existing tolerance does not affect the burden of persuasion established by Congress. . . . Once new evidence bearing on the safety of pesticide residues has been adduced or cited sufficient to justify reopening the issue of the validity of existing tolerances, as in the present case, the burden of establishing the safety of any tolerance remains on those who seek to permit a residue."

The second petition, against the Department of Agriculture, called for "suspension," not "cancellation" of DDT, since the latter merely postponed a ban on the chemical until appeals by manufacturers or users had been processed or litigated. (See Chapter 9.) The Department of Agriculture had argued that the petitioners had no standing, that the agency had not yet made a decision on DDT so there was no case for review, and that, anyway, its administrative ruling was not subject to review because of the Department's discretionary authority. These challenges were all rejected and the court, whose opinion was written by Judge David Bazelon, gave the Secretary of Agriculture thirty days to show why it had not suspended DDT. Moreover, wrote Bazelon, "If he persists in denying suspension in the face of the impressive evidence presented by petitioners, then the basis for that decision should appear clearly on the record, not in conclusory terms but in sufficient detail to permit prompt and effective review."

HUDSON RIVER AND THE CORPS

Since Airlie House, conservation defenders have been successful beyond the expectations of that meeting in discovering

action-forcing mechanisms in the government statutes. The
DDT and pipeline decisions provided new hope. And yet
another victory was chalked up in the case of Citizens' Committee for the Hudson Valley v. Volpe. The State of New York
was all set to begin pushing some 9.5 million cubic yards of
fill dirt into the Hudson River, as far out as 1,300 feet, in
order to accommodate a six-lane highway overlapping the course
of the river for about four miles. Since the project involved a
navigable waterway, the state duly had obtained permission
from the Corps of Engineers to fill the river. David Sive, by
now steeped in both the lore and laws of rivers, argued that
Section 401 of the River and Harbor Act forbids the building
of bridges, dams, dikes, or causeways in navigable waterways
without permission from Congress as well as the Corps. Moreover, since authority for bridges and causeways has been transferred to the Department of Transportation, that agency's permission was also needed. The court accepted these arguments
and also firmly rejected the challenge that the conservationists
did not have standing. The matter is not dead, because Congress
and The Department of Transportation can as yet give their
approval, but they also have to comply with NEPA, and so the
conservationists have plenty of ammunition for the next round.

As previously noted in this book, both the 1899 Rivers and
Harbors Act and the accompanying Refuse Act give citizens a
good wedge for opening court review of waterfront activities
that threaten the environment. Malcolm Baldwin, in a paper
originally given to Wisconsin law students, contended that these
acts could be broadly applied to environmental degradation on
both the land and water sides of the shoreline. "For the fact is,"
he wrote, "that no federal or state action now on the horizon has
the potential to protect so much of the coastal zone so effectively
as the Corps of Engineers." The definition of "navigable
waters" is so flexible that it has been stretched to include tidal
flats and marshes. As noted in Appendix I, the courts have ruled

that oil and other chemical pollutants are subject to the control of the Refuse Act. Moreover, Baldwin contended that the law's "mandate to the Corps to control any activity that would 'alter or modify . . . the condition' of navigable waters can certainly include shore erosion, nutrient runoff, silitation, and even visual degradation." These levers are further strengthened by the amended (in 1958) Fish and Wildlife Coordination Act that allows the Secretary of Interior to review Corps permits that might affect marine ecosystems. Assuredly, this act would have been cited had conservationists ever decided to bring action against the Corps' deprivation of water for the Everglades National Park, because it says that "wildlife conservation shall receive equal consideration and be coordinated with other features of water-resources development programs."

Air and Water Quality legislation and other pollution laws have become more meaningful as public concern over environmental quality has generated. However, government prosecutors cannot yet be counted upon. The most shameful reluctance has been exhibited by the Department of Justice's Land and Natural Resource Division, which is supposed to institute court actions against polluters as well as defend environmental interests shared by federal agencies. The trouble is that this division also defends the agencies who exploit the environment and so far these interests have held the upper hand. Time after time, in response to suits by conservation organizations, the Justice Department resorts to the old challenges: no standing or no claim. The Department simply has resisted prosecuting complaints under the 1899 Refuse Act that were submitted after this law was publicized by Congressman Henry Reuss. Hopefully, the Nixon administration's proposal for an Environmental Protection Agency will be followed up by reorganization in the Justice Department, giving the environment a position apart from land-development interests. A precedent for this is the addition of a division to handle consumer protection complaints.

SAVING WILDERNESS

Picking out loopholes and inconsistencies in federal laws seems bound to escalate. For some time the Sierra Club's legal counsels have taken this approach toward the preservation of wilderness. Two cases that had not been finally decided at this writing may be supported by the new environmental policy strictures.

The first is Parker v. U.S., which tests the Multiple Use and Sustained Yield Act of 1960. The Sierra Club brought the action when the Forest Service went ahead and leased land for logging in East Meadow Creek adjacent to the Gore Range-Eagle Nest Primitive Area near Vail, Colo. Denver lawyer Tony Ruckel argued that the Forest Service had an obligation to consider the timberland's importance as a wilderness area buffer strip or even as a link in a continuous stretch of wild land. He cited a guideline established under the 1960 act, saying, "It is the policy of Congress that the national forests are established and shall be administered for outdoor recreation, range, timber, watershed, wildlife, and fish purposes." The court granted a preliminary injunction that at least delayed the buzzing of the chain saws.

In the second case, the Sierra Club opposed plans for a huge ski resort in the Mineral King Valley of the Sierra Range. The Club's lawyer, Leland Selna, contended that the terms of the Disney lease had been stretched far beyond the land-use limitations provided under Forest Service regulations and also that an access road to the area would cross nine miles of park in violation of the National Park Act. This act, argued the Sierra Club attorney, says that developments (e.g., roads and other facilities) in national parks must benefit public enjoy-

ment of the park and be "consistent to the highest practicable
degree with the preservation and conservation of the areas."
Again the Sierra Club won a preliminary injunction and the
right to have the issues debated openly in a court trial, despite
the government's usual contention that there was no standing.

GETTING HELP

While there would appear to be ample leverage for securing
legal support where and when the federal government is even
remotely concerned, there is not yet a law allowing individuals
to sue private polluters. To be sure, you can complain to an
appropriate state or federal agency about such culprits, but you
then must stand idly by and hope that the government will
prosecute your complaint vigorously. As this chapter was written, several proposals were pending in Congress to give the
ordinary citizen much more strength. California Congressman
Richard T. Hanna had proposed an amendment to NEPA and
Senator Thomas Hart had proposed an Environmental Protection Act.

Class Actions and Stockholder Suits

Originally, the Nixon administration's proposals for strength-
ening consumer protection included unlimited provisions for
class action suits (when a citizen complains on behalf of all
those aggrieved or deprived by the activity or party that is
sued), but these were watered down considerably before they
left the White House for Capitol Hill, reportedly much to the
chagrin of the President's adviser on consumer affairs, Virginia
Knauer.

During the spring of 1970, "Campaign GM," a lawyer-led crusade to force General Motors to become more responsive socially and environmentally, opened up new possibilities for stockholders who feel company directors have dictated the use of profits in flagrant disregard of the public interest. The strategy was to acquire enough voting power through proxies and outright support of institutional and individual shareholders to vote new, sympathetic directors to the GM board. While the plan did not succeed, against terrific odds of course, it will get another chance next year and already has strong support. John D. Rockefeller, IV, West Virginia's dynamic young Secretary of State, commented at a Washington press conference that "Campaign GM is a logical outgrowth of something very good that is happening in this country—an increasing demand for accountability and participation."

There are quite a few possibilities for stockholders' suits against corporate irresponsibility toward the environment. The "Campaign GM" approach will undoubtedly be copied. Another tack is to force the company to include in proxy statement mailings resolutions, insisting upon ecological considerations in the firm's operations, that can be voted upon at the annual meeting.

Finding a Lawyer

How do you go about finding a lawyer who is aware not only of the opportunities presented in the foregoing pages but has undoubtedly been experimenting with other approaches? As a result of the Airlie House meeting, the proceedings of which are available (see Bibliography), lawyers everywhere have an opportunity to become knowledgeable through an Environmental Law Reporter and an Environmental Law Institute (see Bibliography). Malcolm Baldwin, who helped organize these new activities, suggests that citizens get legal advice by contacting local units of the American Bar Association or American Trial Lawyers Association. Ask them who has represented environmental litigants in your area. You can also approach the local or regional representative of such conservation organizations as the Sierra Club, the National Audubon Society, and Izaak Walton League. If you are taking on a polluter who has a national base (e.g., a subsidiary of a major industrial corporation), contact the national headquarters of these conservation organizations, write to the Environmental Defense Fund, or even ask for suggestions through your Congressman and Senators.

Costs

Environmental litigation won't come cheap, because rich and powerful interests invariably are being sued. At Airlie House, Joseph Sax noted that to receive a decent salary and pay off expenses a lawyer would have to gross $25,000 for six months or twice that for a year. It was also noted that while conservationists command an enthusiastic following and can obtain witnesses without expense to testify on general issues, scientific experts are increasingly necessary to rebut challenges of corporation and government researchers. Such scientific witnesses can cost between $300 and $700 a day. The daily court transcripts, needed to brief witnesses, cost around $250. Depositions are also expensive, around $1.50 a page. On top of all this, in some cases (e.g., Parker v. U.S.), the government successfully has demanded that a bond be posted by the litigant to indicate that its intentions are serious.

All considered, you are still at a disadvantage taking enemies of the environment to court. Only in one respect do you *appear to always hold an edge*. The cause is righteous—but shouldn't be made to seem too much so—and its banner seems to be attracting more and more followers and greater public understanding.

APPENDIX I

ENFORCEMENT OF 1899 REFUSE ACT THROUGH CITIZEN ACTION

I. *What is Prohibited and Where*—The 1899 Refuse Act is a powerful, but little used, weapon in our Federal arsenal of water pollution control enforcement legislation. Section 13 of the Act (Title 33, United States Code, section 407) prohibits *anyone,* including any individual, corporation, municipality, or group, from throwing, discharging, or depositing any refuse matter of any kind or any type from a vessel or from a shore-based building, structure, or facility into either (a) the Nation's navigable lakes, rivers, streams, and other navigable bodies of water, or (b) any tributary to such waters, unless he has first obtained a permit to do so. Navigable water includes water sufficient to float a boat or log at high water. This section of the Act applies to inland waters, coastal waters, and waters that flow across the boundaries of the United States and Canada and Mexico.

The term "refuse" has been broadly defined by the Supreme Court to include all foreign substances and pollutants. It includes solids, oils, chemicals, and other liquid pollutants. The only materials excepted from this general prohibition are those flowing from streets, such as from storm sewers, and from municipal sewers, which pass into the waterway in liquid form.

In addition, the section prohibits anyone from placing on the bank of any navigable waterway, or of any tributary to such waterway, any material that could be washed into a waterway by ordinary or high water, or by storms or floods, or otherwise and would result in the obstruction of navigation.

II. *Permits to Discharge*—Section 13 of the Act authorizes the Secretary of the Army, acting through the Corps of Engineers, to permit the deposit of material into navigable waters under conditions prescribed by him. Regulations governing the issuance of permits are published in Title 33 of the Code of Federal Regulations, Part 209.

III. *Penalty for Violations*—Violations of the Refuse Act are subject to

criminal prosecution and penalties of a fine of not more than $2500 nor less than
$500 for each day or instance of violation, or imprisonment for not less than 30
days nor more than 1 year, or both a fine and imprisonment (Title 33, United
States Code, Section 411). A citizen, who informs the appropriate United States
attorney about a violation and gives sufficient information to lead to conviction,
is entitled to one-half of the fine set by the court. (See section V of this
outline.)

IV. *Procedure for Citizen to Seek Enforcement of Refuse Act—*
A. The citizen having information about any discharge of refuse into navigable
waters should first ascertain whether the discharge is authorized by Corps permit.
If a permit is in effect, the citizen should endeavor to ascertain whether the
permittee is complying with its terms. This information can be obtained from
the appropriate office of the Corps of Engineers with jurisdiction over the particular waters into which the discharge occurs. Such information is available to
the public under the Freedom of Information Act (5 U.S. Code 552; Public
Law 90-23).

B. The Refuse Act specifically directs that the appropriate United States attorney
shall "vigorously prosecute all offenders." (Title 33, United States Code, section
413.) In order to do so he needs adequate information to prove that the discharges were made and that they violated the law or the conditions of the
permit. Furthermore, the statute specifies that the citizen's right to one-half of
the fine is conditioned on his providing to the U.S. attorney information sufficient to *lead to a conviction of the violator.*

In providing information to the U.S. Attorney, the citizen should make a
detailed statement, sworn to before a notary or other officer authorized to administer oaths, setting forth:

(a) the nature of the refuse material discharged;
(b) the source and method of discharge;
(c) the location, name, and address of the person or persons causing or contributing to the discharge;
(d) the name of the waterway into which the discharge occurred;
(e) each date on which the discharge occurred;
(f) the names and addresses of all persons known to you, including yourself, who saw or knows about the discharges and could testify about them if
necessary;
(g) a statement that the discharge is not authorized by Corps permit, or, if
a permit was granted, state facts showing that the alleged violator is not complying with any condition of the permit;
(h) if the waterway into which the discharge occurred is not commonly
known as navigable, or as a tributary to a navigable waterway, state facts to
show such status;
(i) where possible, photographs should be taken, and samples of the pollutant or foreign substance collected in a clean jar which is then sealed. These

should be labeled with information showing who took the photograph or sample, where, and when, and how; and who retained custody of the film or jar.

Where the material is liable to be washed into the waterway from its bank, in violation of the Act, similar information should also be provided to the U.S. Attorney in such a statement.

C.   When a citizen furnishes information to the U.S. Attorney for the purpose of aiding in the prosecution of violators of the Refuse Act for past discharges, the citizen should also urge the U.S. Attorney to seek injunctions under the same Act to preclude future discharges, or other orders to require the dischargers to remove pollutants already discharged. More frequent use of this authority by the government, together with criminal sanctions, will have lasting pollution control results.

V.   *Qui Tam Suits*—Where a statute, such as the Refuse Act, provides that part of a fine shall be paid to citizens who furnish sufficient information of a violation to lead to a conviction of the violator, and the government fails to prosecute within a reasonable period of time, the informer can bring his own suit, in the name of the government, against the violator to collect his portion of the penalty. This is called a *qui tam* suit. The informer has a financial interest in the fine and therefore can sue to collect it. The Supreme Court has upheld such *qui tam* suits. Some of these decisions are cited in the Report of the House Committee on Government Operations (House Report 91-917, March 18, 1970) entitled "Our Waters and Wetlands: How the Corps of Engineers Can Help Prevent Their Destruction and Pollution."

The United States district courts apparently have exclusive jurisdiction to hear and decide such suits. (Title 28, United States Code, section 1355.) In such a *qui tam* suit, the citizen must prove that the alleged violator did, in fact, violate the Act.

If the citizen should lose his suit, he probably would have to bear the cost of suing, including his lawyer's fees.

APPENDIX II
AGENCIES AND
ORGANIZATIONS

THE FOLLOWING WAS COMPILED BY THE CITIZENS' ADVISORY
COMMITTEE ON ENVIRONMENTAL QUALITY. THE AUTHOR HAS
ADDED TO THE LIST SEVERAL OTHER ORGANIZATIONS.

FEDERAL AGENCIES

Among the operating agencies of the federal government there are many
that provide grant-in-aid programs for local action, technical assistance, education, and research services. They also publish many helpful guides and pamphlets
and are usually delighted when citizens write and ask for them. In this appendix,
we can note only the principal services offered; for a more complete listing you
should send for "Catalog of Federal Domestic Assistance." This is available
free from the Information Center, Office of Economic Opportunity, Washington,
D.C. 20506.

Generally, you will save time by writing to federal agencies at their regional
offices rather than the Washington office, and for that reason, we have provided
a geographic listing of local offices of key agencies.

Bureau of Outdoor Recreation

Department of the Interior, Washington, D.C. 20240. The Bureau coordinates
federal recreation programs and administers matching grants to states for state
and local outdoor recreation planning, land acquisition, and development projects.
It can advise on a wide range of problems involved in state, county, and regional
outdoor recreation programs.

REGIONAL OFFICES

NORTHEAST	MID CONTINENT
1421 Cherry Street	Denver Federal Center, Bldg. 41
Philadelphia, Pa. 19102	Denver, Colo. 80225

SOUTHEAST	PACIFIC NORTHWEST
810 New Walton Building	U.S. Courthouse, Room 407
Atlanta, Ga. 30303	Seattle, Wash. 98104
LAKE CENTRAL	PACIFIC SOUTHWEST
3853 Research Park Drive	450 Golden Gate Avenue, Box 36062
Ann Arbor, Mich. 48104	San Francisco, Cal. 94102

Federal Water Pollution Control Administration

Department of the Interior, 633 Indiana Avenue, N.W., Washington, D.C. 20240. It makes grants for comprehensive river basin planning, for the construction of waste treatment work, and for research, development, and demonstration projects. It conducts research, provides technical assistance, and carries out training programs.

REGIONAL OFFICES

NORTHEAST REGION

(Conn., Del., Me., Mass., N.H.,
N.J., N.Y., R.I., Vt.)
John F. Kennedy Building, Room 2303
Boston, Mass. 02203

MIDDLE ATLANTIC REGION

(D.C., Md., N.C., Pa., S.C., Va.)
918 Emmet Street
Charlottesville, Va. 22901

SOUTHEAST REGION

(Ala., Fla., Ga., Miss., P.R.,
Tenn., Virgin Islands)
1421 Peachtree Street, N.E., Suite 300
Atlanta, Ga. 30309

SOUTH CENTRAL REGION

(Ark., La., N.M., Okla., Tex.)
1402 Elm Street
Dallas, Tex. 75202

OHIO BASIN REGION

(Ind., Ky., O., W. Va.)
4676 Columbia Parkway
Cincinnati, O. 45226

GREAT LAKES REGION

(Ill., Ia., Mich., Minn., Wisc.)
33 East Congress Parkway, Room 410
Chicago, Illinois 60605

MISSOURI BASIN REGION

(Colo., Kan., Mo., Nebr., N.D.,
S.D., Wyo.)
911 Walnut St., Room 702
Kansas City, Mo. 64106

SOUTHWEST REGION

(Ariz., Calif., Hawaii, Nev.,
Utah, Guam)
760 Market St.
San Francisco, Calif. 94102

NORTHWEST REGION

(Alas., Idaho, Mont., Oreg., Wash.)
Room 570, Pittock Block
Portland, Ore. 97205

The Department of Agriculture

Washington, D.C. 20250. Its many agencies have become increasingly involved in programs for recreation and landscape conservation, and in urban as well as rural areas. The Soil Conservation Service provides technical and financial assistance through local soil and water conservation districts, and as part of resource projects and small watershed flood control projects. The Farmers Home Administration can provide credit assistance to farmers for development on income-producing recreation enterprises. The Agricultural Stabilization and Conservation Service can help share costs for special conservation projects benefiting the community. The Forest Service provides guidance for recreational development of private woodlands.

The Department has a field man for one or more of its agencies in every county, and you should go to him to find out more about the various programs. Agencies will be listed in phone directories under U.S. Government—Agriculture, or County Extension Agent.

Department of Health, Education and Welfare

Washington, D.C. 20201. The Office of Education provides Title III grants and services for environmental education programs, facilities, and materials. The Public Health Service provides research, training, technical assistance, and grants-in-aid for air pollution and solid waste control.

REGIONAL OFFICES

REGION I

(Conn., Me., Mass., N.H., R.I., Vt.)
John F. Kennedy Federal Building
Government Center
Boston, Mass. 02203

REGION II

(N.J., N.Y., P.R., Virgin Islands)
Federal Building
26 Federal Plaza
New York, N.Y. 10007

REGION III

(Pa., Del., Md., D.C., Va., W.Va.)
220 Seventh St., N.E.
Charlottesville, Va.

REGION IV

(N.C., S.C., Ky., Tenn., Miss., Ala., Ga., Fla.)
50 Seventh Street, NE
Atlanta, Ga. 30323

REGION V

(Ill., Ind., Mich., Minn., O., Wisc.)
433 West Van Buren St.
Chicago, Ill. 60607

REGION VI

(Ark., La., N.M., Okla., Tex.)
1114 Commerce St.
Dallas, Tex.

REGION VII

(Ia., Kan., Mo., Neb.)
601 East 12th St.
Kansas City, Mo. 64106

Department of Housing and Urban Development

Washington, D.C. 20410. It provides grants in metropolitan areas to expand community beautification programs, to help state and local governments acquire open space, prepare comprehensive local, regional, or statewide plans (including open space and outdoor recreation plans).

REGIONAL OFFICES

REGION I

(Conn., Me., Mass, N.Y., N.H., R.I., Vt.)
26 Federal Plaza
New York, N. Y. 10007

REGION II

(Del., D.C., Md., N.J., Pa., Va., W. Va.)
6th & Walnut St.
Philadelphia, Pa. 19106

REGION III

(Ala., Fla., Ga., Ky., Miss., N.C., S.C., Tenn.)
645 Peachtree-Seventh Building
Atlanta, Ga. 30323

REGION IV

(Ill., Ind., Ia., Mich., Minn., Neb., N.D., O., S.D., Wis.)
360 N. Michigan Ave., Room 1500
Chicago, Ill. 60601

REGION V

(Ark., Colo., Kan., La., Mo., N.M., Okla., Tex.)
819 Taylor Street
Fort Worth, Texas 76102

REGION VI

(Ariz., Calif., Guam, Hawaii, Nev., Southern Idaho, Utah, Wyo.)
450 Golden Gate Ave.
San Francisco, Calif. 94102
(Alaska, Mont., Northern Idaho, Ore., Wash.)
Area Office: 2nd & Union
Seattle, Wash. 98101

REGION VII

(P.R. and the Virgin Islands)
P.O. Box 3869
GOP San Juan, P.R. 00936

STATE AGENCIES

State agencies can be very useful; like federal agencies, most of them have extensive aid, and educational and publication programs, and an inquiry directed to them about available material will usually receive prompt attention. In many cases they have staff men whose job it is to work with local groups and with landowners.

The names of the agencies vary from state to state; in one, for example, the principal conservation agency might be called the Department of Natural Resources; in another, the Conservation Commission. However, if you write to the Department of Conservation and Natural Resources, State Capital, your letter will probably find its way to the right office. Similarly, a letter directed to the Department of Parks and Recreation will reach the principal recreation agency, whatever its precise title. State planning agencies can be key contacts too.

Most state agencies have local offices in the principal cities, and for a condensed list of their names, addresses, and phone numbers there is nothing to beat the phone directory. Under the main listing for the state will be a list of the principal agencies. It is also a good idea to see if there is any listing under the particular subject you are concerned with. Look up "air" or "water" or "park" for example, and you may run across additional agencies and groups that would be of help. The phone directory is probably the single most important tool for launching campaigns, but it is amazing how many people overlook the usefulness of it as a source of leads and instantaneous information.

The State University is another source of help you should inquire about. Increasingly, State Universities are doing advisory and research work in environmental resource problems. Virtually every State University has an agricultural extension service, and many have services for urban problems.

PRIVATE ORGANIZATIONS

Most of the organizations listed here provide informational and publication services and a number have staff people to lend guidance to local groups. Many of these organizations have branches or charters in the states and cities. For a more complete listing of organizations in the conservation field, you should consult the "Conservation Directory." This is published annually by the National Wildlife Federation, 1412 16th Street, N.W., Washington, D.C. 20036; $1.50 a copy. It is a useful guide for finding allies.

THE CONSERVATION FOUNDATION

1250 Connecticut Avenue, N.W., Washington, D.C. 20036. Through research the Foundation seeks to further knowledge about the interaction between man and nature; it also seeks to have this knowledge applied to the practical problems of urban growth, such as river basin planning, highway design, and regional development policies. It serves as a clearing house on information about significant new legislation and governmental programs, help for better conservation education in our schools, and has an extensive audio-visual and publications program.

DUCKS UNLIMITED

P.O. Box 66300, Chicago, Ill. 60666. Although its objectives are limited, they are significant. DU has acquired or protected over two million acres of vital breeding habitats for migrating waterfowl and has ambitious plans for "managing" more wetland areas where the ducks breed in Canada.

ENVIRONMENTAL DEFENSE FUND, INC.

P.O. Box 740, Stony Brook, N.Y. 11790. EDF does not lobby in the Congress, but it goes all out to take on law suits against the enemies of the environment throughout the U.S., from state and federal agencies supporting unsound programs to big industrial polluters.

FRIENDS OF THE EARTH

451 Pacific Avenue, San Francisco, Calif., 94133. This is one group that makes no bones about lobbying vigorously and resorting to legal methods to achieve environmental quality and protection. It was started by former Sierra Club executive director, David Brower.

THE GARDEN CLUB OF AMERICA

598 Madison Avenue, New York, New York 10022. A national organization representing numerous local garden clubs. Active at the local level in beautification, conservation, and open space planning. Distributes a free conservation packet, "The World Around You."

GENERAL FEDERATION OF WOMEN'S CLUBS

1734 N Street, N.W., Washington, D.C. 20036. Unites and serves affiliated local clubs. Its biennial Community Improvement Program offers incentive awards to clubs for outstanding projects to meet local needs, including outdoor recreation needs. Its Conservation Department assists clubs with conservation and outdoor recreation projects.

THE IZAAK WALTON LEAGUE OF AMERICA

1326 Waukegan Road, Glenview, Illinois 60025. A membership organization with local chapters and state divisions; also national memberships. Promotes conservation of renewable natural resources and development and protection of high quality outdoor recreation opportunities. Chapters and divisions can furnish speakers and literature. Publishes monthly Izaak Walton Magazine. Maintains a Conservation Office at 719 13th Street, N.W., Washington, D.C. 20005.

KEEP AMERICA BEAUTIFUL, INC.

99 Park Avenue, New York, New York 10016. A national nonprofit, public service organization for the prevention of litter and for the enhancement of urban and rural scenic and man-made beauty. Publishes helpful brochures and newsletters on litter prevention.

LEAGUE OF WOMEN VOTERS OF THE UNITED STATES

1730 M Street, N.W., Washington, D.C. 20036. A membership organization, with local and state Leagues, dedicated "to promote political responsibility through informed and active participation of citizens in government." Members participate in water resource programs at all levels of government. Many local and state Leagues are interested in open space, parks, and outdoor recreation facilities. Its national office can assist local Leagues in study and action programs.

NATIONAL ASSOCIATION OF COUNTIES

Suite 522, 1001 Connecticut Avenue, N.W., Washington, D.C. 20036. A national, nonprofit membership organization which acts as a clearing house for information relating specifically to county government administration. Publishes a variety of materials relating to parks, air pollution, water pollution, etc.

NATIONAL ASSOCIATION OF SOIL AND WATER CONSERVATION DISTRICTS

1025 Vermont Avenue, N.W., Washington, D.C. 20005. A membership organization of local districts and their state associations through which farmers and other landowners express their views on "judicious use of land, water, timber, and related resources." Its Recreation and Wildlife Committee and local districts can advise landowners considering income-producing recreational enterprises.

NATIONAL AUDUBON SOCIETY

1130 Fifth Ave., New York, N.Y. 10028. A membership organization dedicated to the conservation of wildlife and the natural environment. It has 150 local chapters; operates 40 wildlife sanctuaries across the country. It provides a wide variety of teaching aids to introduce school children to nature study. Its Nature Centers Division has provided guidance in planning and operating community nature centers. Intensive summer programs at four Audubon Camps offer adult courses in ecology for teachers and youth leaders. Publishes two-bimonthlies, *Audubon* magazine and *Audubon Field Notes*.

NATIONAL COUNCIL OF STATE GARDEN CLUBS

4401 Magnolia Avenue, St. Louis, Missouri 63110. The State Clubs conduct a variety of programs for the beautification of the countryside and the cities. They also sponsor adult education courses in landscape principles and techniques.

NATIONAL PARKS ASSOCIATION

1701 18th St., N.W., Washington, D.C. 20009. Dedicated to the protection and acquisition of public park lands, the NPA has widened its sights in recent years and become active in general environmental issues such as water resource management, pesticides, and pollution. It publishes an excellent magazine.

NATIONAL RECREATION AND PARK ASSOCIATION

1700 Pennsylvania Avenue, N.W., Washington, D.C. 20006. A national membership service organization active in the fields of recreation and park development, conservation, and beautification. It has a community service department, publishes many helpful booklets and a monthly magazine, "Parks and Recreation."

NATIONAL TRUST FOR HISTORIC PRESERVATION

748 Jackson Place, N.W., Washington, D.C. Semi-public agency set up to preserve historic properties of national significance and to encourage local preservation efforts. Publishes a quarterly magazine, *Historic Preservation,* and a monthly newspaper, *Preservation News.*

NATIONAL WILDLIFE FEDERATION

1412 16th Street, N.W., Washington, D.C. 20036. Seeks to encourage citizen and governmental action for the conservation of natural resources. Publishes *The Conservation Directory* annually at $1.50 a copy.

NATURAL SCIENCE FOR YOUTH FOUNDATION

763 Silvermine Road, New Canaan, Conn. 06840. Helps communities set up natural science centers, wildlife preserves, and trailside museums for involving young people first hand with the world of nature.

THE NATURE CONSERVANCY

1522 K Street, N.W., Washington, D.C. 20005. A membership organization with state chapters. It seeks to preserve natural areas by direct acquisition and by assistance to educational institutions, private groups, or public agencies. Through its revolving fund it often secures endangered property and holds it for later resale to public agencies.

OPEN SPACE INSTITUTE

145 E. 52nd St., New York, N.Y. 10022. Action group which stimulates open space conservation by working with land owners, municipal agencies, civic and regional groups

POPULATION REFERENCE BUREAU

1775 Massachusetts Ave., N.W., Washington, D.C., 20036. This is the best information center or clearing house for data concerning the effects of the population explosion, in the U.S. and worldwide. PRB regularly publishes bulletins, papers, and bibliographies concerning demographic, economic, and social ramifications of growth.

ROADSIDE COUNCILS

In a number of states, Roadside Councils have been set up to work for highway billboard controls, scenic highways, and roadside rests. There is, unhappily, no national organization, but the California Roadside Council serves as a clearing house for state councils. It is located at 2636 Ocean Drive, San Francisco, California 94132.

SEARS, ROEBUCK FOUNDATION

Chicago, Illinois 60607. Sears has an extensive program for stimulating local action for conservation and beautification. Through the Women's Clubs it gives grants to groups for pace setting projects and each year awards prizes for outstanding accomplishment.

SIERRA CLUB

1050 Mills Tower, San Francisco, California 94104. Devoted to study and protection of the nation's scenic resources—mountains, shorelines, parks, waters, forests, wildlife. It provides films, manuals, exhibits, speakers; sponsors conferences. Its quality book publishing program is unique in conservation.

THE URBAN COALITION

2100 M St. N.W., Washington, D.C. 20037. A nonprofit organization aimed at spurring people and groups to join together in action on the major problems of their cities.

URBAN LAND INSTITUTE

1200 18th Street, N.W., Washington, D.C. 20036. An organization of commercial developers and others interested in planning and development of urban areas. Outstanding for its hard-headed studies of land use patterns.

THE WILDERNESS SOCIETY

729 15th St., N.W., Washington, D.C. 20005. Its main goal is still the protection of wild lands and the acquisition of additional wilderness or primitive areas by the federal government. It has been a significant influence of opposition to oil exploration in the Alaskan Arctic.

YOUNG WOMEN'S CHRISTIAN ASSOCIATION

National Board, 600 Lexington Avenue, New York, New York 10022. Long noted for its work in training young people in the appreciation and use of our outdoor resources, it is also becoming increasingly active in stimulating community planning and beautification efforts through its local chapters. It has an excellent publications program.

APPENDIX III
KEY
LEGISLATIVE
COMMITTEES

tivity as of the summer of 1970. In addition, the appendix ... critical. From *Science* magazine.

COMMITTEE	CHAIRMAN	RANKING MINORITY MEMBER
	Senate	
Agriculture and Forestry	Allen J. Ellender (D-La.)	George D. Aiken (R-Vt.)
Commerce	Warren G. Magnuson (D-Wash.)	Norris Cotton (R-N.H.)
Subcommittee on Energy, Natural Resources, and the Environment	Philip A. Hart (D-Mich.)	Clifford P. Hansen (R-Wyo.)
Government Operations	John L. McClellan (D-Ark.)	Karl E. Mundt (R-S.D.)
Subcommittee on Intergovernmental Relations	Edmund S. Muskie (D-Me.)	Karl E. Mundt (R-S.D.)
Interior and Insular Affairs	Henry M. Jackson (D-Wash.)	Gordon Allot (R-Colo.)
Labor and Public Welfare	Ralph Yarborough (D-Tex.)	Jacob K. Javits (R-N.Y.)
Subcommittee on Health	Ralph Yarborough (D-Tex.)	Peter H. Dominick (R-Colo.)
Public Works	Jennings Randolph (D-W.Va.)	John S. Cooper (R-Ky.)
Subcommittee on Air and Water Pollution	Edmund S. Muskie (D-Me.)	James C. Boggs (R-Del.)
	Joint	
Atomic Energy *	John O. Pastore (D-R.I.)	George D. Aiken (R-Vt.)
	Chet Holifield (D-Calif.)	Craig Hosmer (R-Calif.)
	House of Representatives	
Agriculture	W. R. Poage (D-Tex.)	Page Belcher (R-Okla.)
Government Operations	William L. Dawson (D-Ill.)	Florence P. Dwyer (R-N.J.)
Subcommittee on Conservation and Natural Resources	Henry S. Reuss (D-Wis.)	Guy Vander Jagt (R-Mich.)
Interior and Insular Affairs	Wayne N. Aspinall (D-Colo.)	John P. Saylor (R-Pa.)
Interstate and Foreign Commerce	Harley O. Staggers (D-W.Va.)	William L. Springer (R-Ill.)
Merchant Marine and Fisheries	Edward A. Garmatz (D-Md.)	William S. Mailliard (R-Calif.)
Subcommittee on Fisheries and Wildlife Conservation	John D. Dingell (D-Mich.)	Thomas M. Pelly (R-Wash.)
Subcommittee on Oceanography	Alton Lennon (D-N.C.)	Charles A. Mosher (R-Ohio)
Public Works	George N. Fallon (D-Md.)	William C. Cramer (R-Fla.)
Subcommittee on Flood Control	Robert E. Jones (D-Ala.)	Don H. Clausen (R-Calif.)
Subcommittee on Rivers and Harbors	John A. Blatnik (D-Minn.)	William H. Harsha (R-Ohio)
Science and Astronautics	George P. Miller (D-Calif.)	James G. Fulton (R-Pa.)
Subcommittee on Science, Research, and Development	Emilio Q. Daddario (D-Conn.)	Alphonzo Bell (R-Calif.)

* The chairmanship of this joint committee alternates between the House and the Senate.

BIBLIOGRAPHY

THE SOURCES LISTED in the following pages are not necessarily the most authoritative or encompassing, but if you wish to follow up a particular chapter or become actively involved in the subject, this list will prove helpful. It is material that I have found useful as well as reading which is widely recommended by the experts. Much of my research, aside from data gathered in interviews, comes from obscure documents, transcripts of government hearings or environmental symposia, trial briefs and decisions, task force papers and speeches by many of the people named in this book. It would be impossible to cite all of this information since it is embedded either in my mind or my notebooks.

General References

Arvill, Robert, *Man and Environment.* Penguin Books, 1967.

Dubos, Rene, *So Human an Animal.* Scribner's, 1969.

Environmental Quality, The first annual report to Congress by the President's Council On Environmental Quality, August, 1970.

Faltermayer, Edmund K., *Redoing America.* Harper & Row, 1968.

Farb, Peter, *The Face of North America.* Harper & Row, 1965.

From Sea to Shining Sea, Report by President's Council on Recreation and Natural Beauty, 1968.

Future Environments of North America. Edited by F. Fraser Darling and John P. Milton, Natural History Press, 1966.

Highsmith, Richard M., Jr., J. Granville Jensen, and Robert D. Rudd, *Conservation in the United States.* Rand McNally, 1969.

Institutes for Effective Management of the Environment, Report by Environmental Study Group, National Academy of Sciences, 1970.

Nash, Roderick, ed., *The American Environment.* Selected readings, Addison-Wesley, 1968.

Odum, Eugene P., *Ecology.* Holt, Rinehart & Winston, 1969.

Shepard, Paul, and David McKinley, eds., *The Subversive Science.* Houghton-Mifflin, 1969.

Statistical Abstract of the United States, published annually by U.S. Bureau of the Census.

Storer, John H., *The Web of Life.* The Devin-Adair Co., 1956.

Udall, Stewart, *Agenda for Tomorrow.* Harcourt, Brace & World, 1968.

Chapter I

A Preliminary Study of the Effects of Water Circulation in the San Francisco Bay Estuary, prepared for U.S. Geological Survey, Dept. of Interior, by D. S. McCulloch, D. H. Peterson, P. R. Carlson, and T. J. Conomos, 1970.

Air and Water Pollution Subcommittee of Senate Public Works Committee, *Hearings on Water Quality,* 1969 and 1970.

Clean Water for the 1970's, Federal Water Quality Administration, Dept. of Interior, staff report, 1970.

Clean Water—It's up to You, Izaak Walton League, 1967.

Cleaning Our Environment—The Chemical Basis for Action, Report by American Chemical Society subcommittee on Environmental Improvement, 1969.

Dasmann, Raymond F., *The Destruction of California.* Macmillan, 1965.

Examination into the Effectiveness of the Construction Grant Program for Abating, Controlling and Preventing Water Pollution, Report to Congress by Comptroller General, Nov., 1969.

Leopold, Luna B., and W. B. Langbein, *A Primer on Water.* U.S. Geological Survey, Dept. of Interior, 1960.

Mackenthun, Kenneth M., *The Practice of Water Pollution Biology.* FWQA, Dept. of Interior, 1969.

Pollution Caused Fish Kills, 1968, FWQA, Dept. of Interior, 1969.

Potomac River Water Quality, FWQA Report, Dept. of Interior, 1969.

Sachs, David Peter, "Drink at your Own Risk." *McCalls,* Nov., 1968.

Smith, Anthony Wayne, *A New Water Policy,* testimony of the National Parks Association President before the National Water Commission, Nov. 7, 1969.

The Big Water Fight, League of Women Voters, Stephen Greene Press, Brattleboro, Vt., 1966.

The Cost of Clean Water and its Economic Impact, FWQA Report to Congress, 1970.

Water Quality Criteria, Dept. of Interior, 1968.

Who Pays For a Clean Stream? League of Women Voters, 1969.

Chapter II

Air Pollution, report by Scientists Institute for Public Information, 1970.

Ayres, Robert U., *Two Possible Alternatives to the Internal Combustion Engine.* Resources for the Future, Annual Report, 1968.

Center for Study of Responsive Law, *Task Force Report on Air Pollution*, directed by John C. Esposito, preliminary draft, May, 1970, and since available in paperback, Ballantine.

Cleaning Our Environment, op. cit., American Chemical Society, 1969.

Conservation Foundation, newsletter, *Citizens Play Crucial Role as the Long Fight Against Air Pollution Progresses*, Nov., 1969.

Effect of Chronic Exposure to Low Levels of Carbon Monoxide on Human Health, Behavior and Performance, National Academy of Sciences Report, 1969.

HEW Federal R/D Plan for Unconventional Low Pollution Vehicles, 1969.

HEW Air Quality Criteria for: Sulfur oxides, Particulate Matter, Carbon Monoxide, hydrocarbons and Photochemical oxidants.

Iglauer, Edith, "The Ambient Air." *The New Yorker*, April 13, 1968.

National Emission Standards Study, HEW, 1970.

Progress in the Prevention and Control of Air Pollution, HEW, 1970.

Search for a Low Emission Vehicle, Senate Commerce Committee Report, 1969.

Chapter III

Baron, Robert A., *The Tyranny of Noise*. St. Martin's Press, 1969.

Beranek, Leo, *Noise, Health and Architecture*, paper of June, 1969, Bolt, Beranek and Newman, Inc.

Cohen, Dr. Alexander, Bureau of Occupational Safety and Health, HEW, Testimony before Senate Subcommittee on Air and Water Pollution, March, 1970.

Conservation Foundation, Newsletter, *An Opening Attack on the Decibel Din*, August, 1968.

Mayor Lindsay's Task Force on Noise, 1970.

Mecklin, John M., "It's Time to Turn Down All That Noise." *Fortune*, Oct., 1969.

Noise—A New Focus for Government and Industry, conference in Washington, Feb., 1969, sponsored by the National Council on Noise Abatement.

Noise: Sound Without Value, Committee on Environmental Quality of Federal Council for Science and Technology, Sept., 1968.

Shurcliff, William A., *SST and Sonic Boom Handbook*. Ballantine, 1970.

Chapter IV

Abrahamson, Dean E., *Environment Costs of Electric Power*. Scientists Institute for Public Information, 1970.

Borgstrom, Dr. Georg A., *The Dual Challenge of Health and Hunger—
A Global Crisis.* Population Reference Bureau, Inc., Jan., 1970.
*Effects of Population Growth on Natural Resources and the Environ-
ment,* Hearings before House subcommittee on Conservation and
Natural Resources, Sept., 1969.
Ehrlich, Dr. Paul R., *The Population Bomb.* Ballantine, 1968.
Hardin, Garrett, *Population, Environment and Birth Control.* W. H.
Freeman, 1969.
Mayer, Jean, *Toward a Non-Malthusian Population Policy.* Columbia
Forum, Summer, 1969.
Miles, Rufus E., Jr., *Whose Baby Is the Population Problem?* Popula-
tion Reference Bureau, Feb., 1970.
Mishan, Ezra J., *Costs of Economic Growth.* Praeger, 1967.
Resources and Man, report for National Academy of Sciences, W. H.
Freeman, 1969.
Udall, Stewart, *The Quiet Crisis.* Holt, Rinehart & Winston, 1963.

Chapter V
*A Comprehensive Assessment of Solid Waste Problems, Practices and
Needs,* Report for President by Stanford Research Institute, 1968.
An Interim Report, *1968 National Survey of Community Waste Prac-
tices,* Bureau of Solid Waste Management, HEW, 1969.
Cleaning Our Environment, op. cit., American Chemical Society.
Darnay, Arsen, and William E. Franklin, *The Role of Packaging in
Solid Waste Management, 1966 to 1976,* Report for HEW, 1969.
Jensen, Michael E., *Observations of Continental European Solid Waste
Management Practices,* Report for HEW, 1969.
Policies for Solid Waste Management, National Academy of Sciences
Report for HEW, 1970.
Senate Subcommittee on Air and Water Polllution, Hearings, 1968, on
Waste Management Research and Environmental Quality and, in
1969 and 1970 on *The Resource Recovery Act of 1969.*
Solid Waste Management, a ten-pamphlet study with recommendations,
by the National Association of Counties Research Foundation,
1969.
Summaries of Solid Wastes Demonstration Grant Projects, 1968 and
1969, Bureau Solid Waste Management.
Waste Management and Control, NAS Report for Federal Council for
Science and Technology, 1966.

Chapter VI
The National Audubon Society, The Sierra Club, Environmental

Action in Washington, and other conservation organizations, nationally or through local chapters, have printed check-lists. Otherwise, it's up to you to make up your own list of good habits.

Chapter VII

American Institute of Planners, *New Communities: Challenge for Today,* edited by Muriel I. Allen, 1968.

Dasmann, Raymond F., *A Different Kind of Country.* Macmillan, 1968.

Halprin, Lawrence E., *Cities.* Reinhold, 1963.

Jacobs, Jane, *The Economy of Cities.* Random House, 1969.

Leopold, Aldo, *A Sand County Almanac.* Oxford, 1949.

Little, Charles E., *Challenge of the Land.* Pergamon Press, 1969.

McHarg, Ian, *Design with Nature.* Natural History Press (Doubleday), 1969.

McQuade, Walter, "Urban Expansion Takes to the Water." *Fortune,* Sept., 1969.

Mumford, Lewis, *The City in History.* Harcourt, Brace & World, 1961

Mumford, Lewis, *The Urban Prospect.* Harcourt, Brace & World, 1968.

National Commission on Urban Problems, chaired by former Sen. Paul Douglas, *Zoning Controversies in the Suburbs,* Research Report 11, 1968; *U.S. Land Prices—Directions and Dynamics,* Report 13, 1968; *Fragmentation in Land Use Planning and Control,* Report 18, 1969.

Open Space Institute, *Stewardship,* 1965, and *Operations Manual for a Local Landowner Program,* 1970.

Osborn, Frederic J., and Arnold Whittick, *The New Towns: The Answer to Megalopolis.* Leonard Hill Books, London, 1969.

Public Land Law Review Commission, Report to Congress, *One Third of the Nation's Land,* 1970.

Roueche, Berton, *What's Left: Report on a Diminishing America.* Little, Brown & Co., 1969.

Scheffey, Andrew J. W., *Conservation Commissions in Massachusetts,* with supplementary report, *Idea on the Move,* by William J. Duddleson, Conservation Foundation, 1969.

Strong, Ann Louise, *Open Space for Urban America,* study for Urban Renewal Administration, HUD, 1965.

Surface Mining and Our Environment, Dept. of Interior, 1967.

Where Not to Build, Technical Bulletin of Bureau of Land Management, Dept. of Interior, 1968.

Whyte, William H., *The Last Landscape.* Anchor, 1970.

Wood, Samuel E., and Alfred E. Heller, *California Going, Going,*

1962; *The Phantom Cities*, 1963, California Tomorrow, San Francisco.

Von Eckhardt, Wolf, *A Place To Live*. Delta, 1967.

Chapter VIII

Conservation Foundation, Newsletters, *Needed: Effective Management of Our Priceless Shorelines and Estuaries*, May 1970; and *A Classic Confrontation in California: Citizens Move to Save San Francisco Bay*, June, 1969.

Estuarine Areas, Hearings before the House Sub-committee on Fisheries and Wildlife Conservation, 1967.

Johnson, Peter L., *Wetlands Preservation*. Open Space Institute, 1969.

Marx, Wesley, *The Frail Ocean*. Ballantine, 1967.

National Estuarine Pollution Study, FWQA, Dept. of Interior, 1969.

National Estuarine Study, Bureau of Sport Fisheries and Wildlife, Dept. of Interior, 1970.

Oil Pollution, Report to the President by Depts. of Interior and Transportation, 1968.

Our Nation and the Sea, Commission on Marine Sciences Report to President, 1968.

Our Wetlands: How the Corps of Engineers Can Help Prevent Their Destruction and Pollution, House Committee on Government Operations Report, 1970.

The Oil Spill Problem and *Offshore Mineral Resources*, Vols. I and II of report by President's Panel on Oil Spills for the Office of Science and Technology, 1969.

Warner, Richard E., *Environmental Effects of Oil Pollution in Canada*. Brief prepared for Canadian Wildlife Service, 1969.

Where Rivers Meet the Sea, League of Women Voters pamphlet, Feb., 1970.

Chapter IX

Carson, Rachel, *The Silent Spring*. Houghton-Mifflin, 1962.

Conservation Foundation, *Pollution by Pesticides*, updated report based on April and May, 1969, newsletters.

Graham, Frank, *Since Silent Spring*. Houghton-Mifflin, 1970.

Knipling, Edward F., *Alternate Methods of Controlling Insect Pests*, paper for Food and Drug Administration, Dept. of Agriculture, 1969.

Pesticides, Scientists' Institute for Public Information, 1970.

Report of the Secretary's Commission on Pesticides and Their Relation-

ship to Environmental Health, chaired by Emil M. Mrak, HEW, Nov., 1969.

Rudd, Robert, *Pesticides and the Living Landscape.* University of Wisconsin, Madison, 1964.

Stickel, Lucille F., *Organochlorine Pesticides in the Environment,* Report for Sport Fisheries and Wildlife, Dept. of Interior, Oct., 1968.

The Poison-Free Garden, special issue of *Cry California,* Summer, 1969, California Tomorrow, San Francisco.

Whiteside, Thomas, "A Reporter at Large." *The New Yorker,* Feb. 7 and March 14, 1970; and "Department of Amplification." *The New Yorker,* June 20 and July 4, 1970.

Chapter X

The written sources for this chapter, cited frequently, are mostly legal briefs and opinions. You can get information on such legal developments by inquiring of The Environmental Law Institute, Dupont Circle Building, Suite 608, 1346 Connecticut Ave. N.W., Washington, D.C., 20036. The institute has just begun publication of a monthly *Environmental Law Reporter.* Otherwise, the following two new books should be most helpful.

Baldwin, Malcolm, and James K. Page, Jr., eds., *Law and the Environment.* Walker, 1970. The proceedings and papers of the Airlie House Conference on the title subject, Sept., 1969.

Sax, Joseph, *Those Who Know Best.* Alfred A. Knopf, 1970.

❧ INDEX

DATE DUE

APR 25 '72	APR 25 '72		
MAY 16 '72	MAY 26 '72		
MAY 15 '78	MAY 22 '78		
APR 18 '80	APR 14 '80		
GAYLORD			PRINTED IN U.S.A.